Springer Theses

Recognizing Outstanding Ph.D. Research

Aims and Scope

The series "Springer Theses" brings together a selection of the very best Ph.D. theses from around the world and across the physical sciences. Nominated and endorsed by two recognized specialists, each published volume has been selected for its scientific excellence and the high impact of its contents for the pertinent field of research. For greater accessibility to non-specialists, the published versions include an extended introduction, as well as a foreword by the student's supervisor explaining the special relevance of the work for the field. As a whole, the series will provide a valuable resource both for newcomers to the research fields described, and for other scientists seeking detailed background information on special questions. Finally, it provides an accredited documentation of the valuable contributions made by today's younger generation of scientists.

Theses may be nominated for publication in this series by heads of department at internationally leading universities or institutes and should fulfill all of the following criteria

- They must be written in good English.
- The topic should fall within the confines of Chemistry, Physics, Earth Sciences, Engineering and related interdisciplinary fields such as Materials, Nanoscience, Chemical Engineering, Complex Systems and Biophysics.
- The work reported in the thesis must represent a significant scientific advance.
- If the thesis includes previously published material, permission to reproduce this must be gained from the respective copyright holder (a maximum 30% of the thesis should be a verbatim reproduction from the author's previous publications).
- They must have been examined and passed during the 12 months prior to nomination.
- Each thesis should include a foreword by the supervisor outlining the significance of its content.
- The theses should have a clearly defined structure including an introduction accessible to new PhD students and scientists not expert in the relevant field.

Indexed by zbMATH.

Tianyi Guo

Low Energy Photon Detection

Doctoral Thesis accepted by the University of Central Florida, USA

Springer

Tianyi Guo
Department of Physics
University of Central Florida
Orlando, FL, USA

ISSN 2190-5053 ISSN 2190-5061 (electronic)
Springer Theses
ISBN 978-3-031-71543-3 ISBN 978-3-031-71544-0 (eBook)
https://doi.org/10.1007/978-3-031-71544-0

This Springer imprint is published by the registered company Springer Nature Switzerland AG
The registered company address is: Gewerbestrasse 11, 6330 Cham, Switzerland

Foreword

Detection of long wave infrared (LWIR) light at room temperature has posed a persistent challenge due to the low energy of photons. The pursuit of an affordable, high-performance LWIR camera capable of room temperature detection has spanned several decades. In the realm of contemporary LWIR detectors, they can be broadly classified into two categories: cooled and uncooled detectors. Cooled detectors, such as MCT detectors, excel in terms of high detectivity and fast response times. However, their reliance on cryogenic cooling significantly escalates their cost and restricts their practical applications. In contrast, uncooled detectors, like microbolometers, are capable of functioning at room temperature and come at a relatively lower cost. But they exhibit lower sensitivity and slower response times.

Within the scope of Dr. Tianyi Guo's PhD work, Dr. Guo showcased two innovative approaches aimed at advancing the next generation of LWIR detectors/cameras. These approaches are designed to offer high detectivity, swift response times, and room temperature operation. The first approach involves harnessing Dirac plasmon and the Seebeck effect in graphene to create a photo-thermoelectric detector. The second approach entails the use of an oscillating circuit integrated with phase change materials and the modulation of frequency induced by infrared illumination to achieve LWIR detection. Finally, Dr. Guo integrated the graphene-based detectors to readout circuits to enable the development of dense pixel focal plane array in collaboration with world's largest IR camera company, Teledyne-FLIR. Such novel material to device actual functional camera development as part of a single PhD is not just unique but unheard off.

Innovations out of the dissertation:

- World's first dynamically tunable light detection
- World's first one-atom thick graphene-based IR camera development
- Discovery of a new photon detection technique based on frequency modulation

Professor
University of Central Florida,
Orlando, FL, USA

Debashis Chanda

Parts of This Thesis Have Been Published in the Following Journal Articles

1. **Frequency Modulation based Long-Wave Infrared Detection and Imaging at Room Temperature**, **Tianyi Guo**, Arindam Dasgupta, Sayan Chandra, Swastik Ballav, et al. *Advanced Functional Materials*, 2309298

Acknowledgments

The work performed herein would not have been possible without the unfailing support of those around me. I wish to thank my advisor, Professor Debashis Chanda, for his contagious passion, drive, and consistent willingness to pursue new projects. It was amazing to feel that our work was limited only by the potential of our ideas and time.

I would like to thank my labmates: Arindam Dasgupta, Pablo Manuel Cencillo-Abad, Sayan Chandra, Muhammad Waqas Shabbir, Alireza Safaei, Aritra Biswas, Manobina Karmakar, and Mahdi Soudi. Many of us have worked together since the beginning and I can't begin to thank you for the half-decade of experiences we've shared. I will always treasure the memories and hope many more will follow.

Lastly, I wish to thank my mother Jing Li. It is tremendously empowering to feel the unconditional love and support you give. I don't want to even imagine a life without you. My whole family was always noticed in the process, from cousins, to aunts, uncles, and my grandparents. I will always remember the caring and love they have given and forever be grateful.

Contents

Abbreviations

A.C.	Alternating Current
CC	Current Control
CVD	Chemical Vapor Deposition
D.C.	Direct Current
EBL	E-Beam Lithography
FDTD	Finite Differential Time Domain
FET	Field Effect Transistor
FWHM	Full Width at Half Maximum
IR	Infrared
LSP	Localized Surface Plasmons
LSPR	Localized Surface Plasmon Resonance
LWIR	Low Wave Infrared Detector
MIT	Metal Insulator Transition
NPG	Nano Patterned Graphene
NW	Nanowire
PCM	Phase Change material
PL	Photolithography
PTE	Photo-Thermoelectric
RIE	Reactive Ion Etching
SEM	Scanning Electron Microscopy
s-SNOM	Scatter-type Scanning Nearfield Optical Microscopy
VC	Voltage Control

Chapter 1
Introduction

Abstract Long-wave infrared (LWIR) detectors, which operate within the 8–12 μm wavelength range, find extensive applications in night vision, space exploration, surveillance, thermal imaging, and defense-related scenarios. Nevertheless, achieving LWIR detection at room temperature presents a formidable challenge due to the inherently low photon energy and the presence of significant background noise. Consequently, the utilization of cryogenic cooling becomes a necessity to attain optimal detectivity for sensitive LWIR detectors. There is a burgeoning demand for the development of efficient room-temperature LWIR detectors, primarily due to their affordability and ease of operation. However, uncooled microbolometers suffer from issues like reduced sensitivity ($D^* \sim 10^8$) and slow response times in the order of tens of milliseconds. As a result, it is crucial to explore new materials and techniques that enable the creation of LWIR detectors operating at room temperature while maintaining exceptional detectivity, achieving fast response times.

1.1 Scope of This Work

The scope of this work is to create a room temperature LWIR detector with higher sensitivity and faster response time compared to commercially available uncooled technologies, while offering spectral tunability and other potential unique properties. This research has also explored the process of upscaling a single pixel created in a laboratory to incorporate sub-million pixels on an electronic readout circuit which is currently used in the industry for prototype camera production (Table 1.1).

T. Guo, *Low Energy Photon Detection*, Springer Theses,
https://doi.org/10.1007/978-3-031-71544-0_1

Table 1.1 Comparison between different LWIR detector technologies

Detector type/properties	Operation condition	Detectivity (Jones)	Response time (ms)
MCT detectors [1]	Cryogenic cooling	~10^{10}–10^{12}	~10^{-5}
Microbolometer [1]	Room temperature	~10^{8}	~7–12
Graphene LWIR detector in this work	Room temperature	~10^{9}	~10^{-4}
Frequency modulation-based detection in this work [2]	Room temperature	~10^{9}	~3

References

1. Tan, C. L., Mohseni, H., et al. (2018). *Nanophotonics, 7*, 169–197. https://doi.org/10.1515/nanoph-2017-0061
2. Guo, T., et al. (2023). Frequency modulation based long-wave infrared detection and imaging at room temperature. *Advanced Functional Materials*, 2309298.

Chapter 2
Dynamically Tunable Long-Wave Infrared Detection

Abstract Detecting long-wave infrared (LWIR) light at room temperature has posed a persistent challenge due to the low energy of photons. The pursuit of an affordable, high-performance LWIR camera capable of room temperature detection has spanned several decades. In the realm of contemporary LWIR detectors, they can be broadly classified into two categories: cooled and uncooled detectors. Cooled detectors, such as Mercury Cadmium Telluride (MCT) detectors, excel in terms of high detectivity and fast response times. However, their reliance on cryogenic cooling significantly escalates their cost and restricts their practical applications. In contrast, uncooled detectors, exemplified by microbolometers, are capable of functioning at room temperature and come at a relatively lower cost. Nonetheless, they exhibit somewhat lower detectivity and slower response times. Within the scope of this work, I will showcase an innovative approach aimed at advancing the next generation of LWIR detectors which utilizes Dirac plasmon and Seebeck effect in graphene.

2.1 Introduction

Long-wave infrared (LWIR) detectors, operating in the 8–14 μm wavelength range, have diverse applications in night vision, space exploration, surveillance, thermal imaging, and defense-related applications [1–5] as shown in Fig. 2.1. However, LWIR detection at room temperature is challenging due to the low photon energy and high background noise. Therefore, cryogenic cooling is required for sensitive LWIR detectors to achieve optimal detectivity [6, 7]. There is growing demand for developing efficient room-temperature LWIR detectors for their low cost and ease of operation. Uncooled microbolometers such as vanadium pentoxide (V_2O_5)- and amorphous silicon (a-Si)-based detectors suffer from lower sensitivity ($D^* \sim 10^8$) and slow response time (10 s of ms) [5, 8, 9]. Furthermore, due to their inherent broadband absorption, both cryocooled detectors and uncooled microbolometers lack the capability of spectral tunability. While the use of tunable IR filters–based LWIR detection can offer spectral tunability [10, 11], their performance is compromised significantly in terms of detectivity. Spectrally tunable LWIR detectors have

T. Guo, *Low Energy Photon Detection*, Springer Theses,
https://doi.org/10.1007/978-3-031-71544-0_2

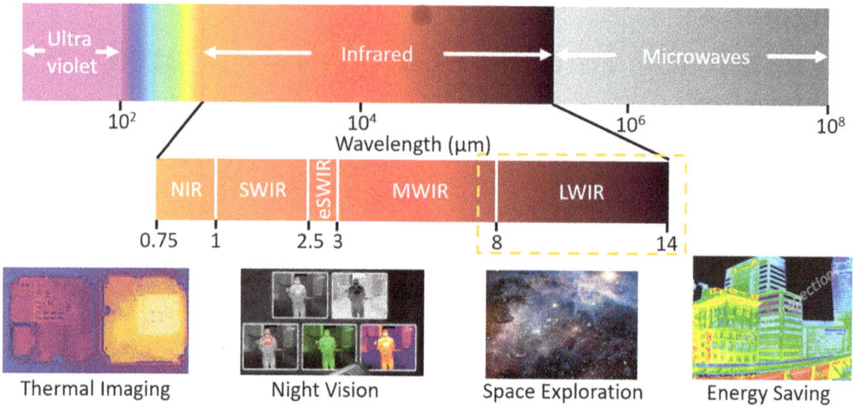

Fig. 2.1 LWIR detection. LWIR detection and its applications in thermal imaging, night vision, space exploration and energy saving

gained huge interest for their potential in various applications such as precise material identification, selective gas sensing, advanced spectroscopic analysis, tailored remote sensing and hyperspectral imaging. Hence, it is imperative to investigate materials and techniques that facilitate the creation of LWIR detectors operating at room temperature with exceptional detectivity, fast response time, and the ability to dynamically tune their operational wavelength.

The emergence of chemical vapor deposition (CVD)-grown large-area graphene has unlocked significant potential for sensitive and fast LWIR detection with its unique Dirac cone band structure, high carrier mobility, and distinctive thermal properties [12, 13]. Furthermore, electrostatic gating enables tailored electronic and optical properties in graphene, allowing for dynamic spectral tunability [14–17]. However, the intrinsic absorption of LWIR radiation in graphene is as low as 2.3% due to the one-atom thickness and low carrier concentration. Prior research has investigated diverse approaches to enhance the light absorption in graphene, including employing additional layers of plasmonic structures and patterning the graphene itself [18, 19]. Patterning graphene directly offers several advantages like seamless integration and increased versatility. Another major challenge in developing an efficient LWIR detector with graphene is achieving effective electrostatic gating for dynamic tunability. The commonly used top or bottom gate geometry has limitations that either obstruct the incident light or prevent the creation of an optical cavity. Instead, a side gate allows direct light access to the graphene layer. In recent years, the use of ion gel as an efficient gate dielectric has enabled the fabrication of large-area graphene-based field-effect transistor (FET) optical devices [20–22]. Ion gel, composed of a polymer matrix infused with an ionic liquid, offers high gate capacitance and ionic conductivity, enabling dynamic control of the Fermi level of graphene over a large area with low gate voltage. A type of ion gel with thin layer thickness, transparency to LWIR radiation, and compatibility with photolithography has been developed; moreover, the stability of such ion gel can be greatly enhanced [23]. Therefore, ion gel gating provides immense potential for developing graphene-based LWIR detectors. While ion gel-gated optical devices with graphene have been demonstrated, there is currently no demonstration of their application in LWIR detection.

In this chapter, I demonstrate a room-temperature LWIR detector utilizing ion gel-gated monolayer graphene, which exhibits dynamically tunable spectral response. By patterning the graphene into a cavity-coupled hexagonal-hole array, we achieve an enhanced absorption of ~70% by exciting Dirac localized plasmons which is further electrostatically tunable within the LWIR spectra range. A carrier temperature gradient is induced in the graphene layer through absorption asymmetry, by patterning half of it with a hexagonal-hole array and leaving the other half unpatterned. The detection relies on photo-thermoelectric effect to convert the asymmetry of carrier temperature into an electrical signal [24–26]. The preliminary detector demonstrates a unique dynamic spectral tuning ability from 8.5 to 11.2 μm while reaching a maximum detectivity of 3.15×10^9 Jones (cryo-cooled detectors ~10^{10} and microbolometers ~10^8) at an ultra-fast response time of 144 ns (cryo-cooled detectors ~μs and microbolometers ~ms), surpassing the capabilities of the conventional uncooled microbolometers by several orders of magnitude. No present cooled or uncooled detectors offer such dynamic spectral tunability and ultrafast response. This demonstration underscores the potential of engineered monolayer graphene LWIR detectors operating at room temperature, offering high sensitivity as well as dynamic spectral tunability for spectroscopic imaging.

2.2 Photo-Thermoelectric Voltage Generation

2.2.1 Plasmon Damping in Graphene

It is essential to gain a comprehensive understanding of the factors that contribute to plasmon losses in graphene. Plasmons within graphene are subject to a range of damping mechanisms, with the primary mode of decay being Landau damping, a process in which a plasmon generates hot carriers as shown in Fig. 2.2. This undesired damping can be alleviated through a process called doping, which elevates the carrier concentration and, in doing so, constrains such transitions over a significant portion of the (q, ω) parameter space due to the Pauli exclusion principle. It's noteworthy that, unlike three-dimensional metallic structures, graphene, being two-dimensional, can be doped using an electrostatic gate voltage. This presents intriguing possibilities for potential applications, as gating provides the means to manipulate plasmon dispersion or even control the activation and deactivation of plasmons.

Furthermore, there exists other plausible mechanisms for damping, including plasmon scattering by various types of impurities or defects, interactions between

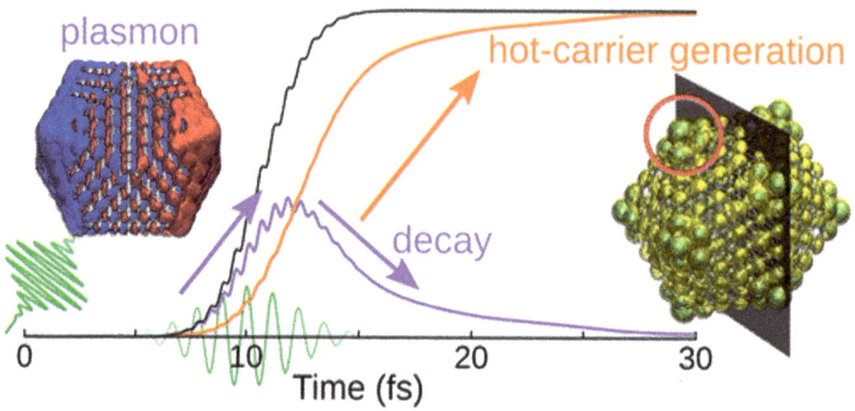

Fig. 2.2 Hot carrier generation through plasmon decay and Landau damping

electrons and phonons, and mutual interactions among electrons. The impact of electron scattering due to impurities can be partially inferred from D.C. measurements, which have revealed exceptionally high electron mobilities, indicative of extended electron lifetimes. As the quality of graphene samples improves, the effects of impurity scattering are anticipated to diminish, thereby posing no significant obstacle to achieving low plasmon losses.

2.3 Experimental Method

2.3.1 Fabrication of the Graphene Infrared Detector

A SiO_2/Si (300 nm/500 μm) substrate was used in the fabrication of the graphene detector. The Au back reflector is first fabricated by photolithography (PL) followed by an E-beam deposition of gold. A 1.3-μm-thick SU-8 layer is then spin-coated and cured on the back reflector to create the Fabry-Perot cavity. A 5-nm-thick Al_2O_3 layer was subsequently deposited on top by atomic layer deposition. Next, a large-area CVD-grown monolayer graphene (Graphenea) is transferred on top of the stack via wet-transfer method. A 10 x 10 μm graphene channel was then formed by PL, reactive ion etching (RIE) with O_2 gas and removal of the photoresist in acetone. The electrodes for three-terminals were formed via PL and E-beam deposition. Next, the hexagonal hole array was patterned on only half of the graphene channel by e-beam lithography and following RIE etching with a radius of 200 nm and a period of 600 nm. Finally, the ion gel was patterned and cured between the graphene and gate electrodes with photolithography by the gelation of a triblock copolymer with an ionic liquid.

2.3.2 Principle of Operation

Upon LWIR illumination, the reflected beam from the bottom Au mirror and the incident beam interferes constructively at the surface of the nano-patterned graphene (NPG) film leading to an enhanced electrical field. The coherent interaction between the optical cavity mode and the Dirac localized plasmons on the patterned graphene significantly enhances LWIR absorption in the NPG layer by about 30 times. A three-terminal field effect transistor (FET) is formed on the graphene with source, drain, and gate electrodes as shown in Fig. 2.3a. An electrical bias to the graphene channel is provided by the source and drain electrodes. The gate electrode is fabricated by the side of graphene. A thin ion gel layer spanning from the gate electrode to the top of NPG functions as the gate dielectric in this side gating arrangement. The carrier concentration in the NPG is dynamically tuned by varying the voltage between the drain and the gate electrodes. Figure 2.3b shows the microscopic image of the fabricated detector with a chemical vapor deposition (CVD)-grown large-area graphene. Details of the device fabrication are included in the Methods section. A scanning electron microscope (SEM) image in Fig. 2.4a shows the half-patterned graphene with circular hole patterns in a hexagonal array. The diameter of the circular holes is 400 nm and the period of the hexagonal array is 600 nm. The proposed detector converts the absorbed light to an electrical signal through Seebeck effect. When exposed to light, localized surface plasmons (LSP) are generated in the patterned side of the graphene layer near the edge of the circular holes. The excited LSP eventually dissipates the energy through damping pathways in graphene. In general, various damping pathways such as phonon emission, bulk scattering, and resistive loss can be prominant [27, 28]. However, in our case, the dominant damping pathway is the edge-assisted Landau scattering, which leads to an increase in electron temperature and the generation of hot carriers in the patterned region of the graphene while the lattice temperature remains unaffected due to the slower lattice heating time. For the unpatterned half of the graphene, it exhibits low absorption to LWIR light of approximately 2%. Therefore, the charge carrier

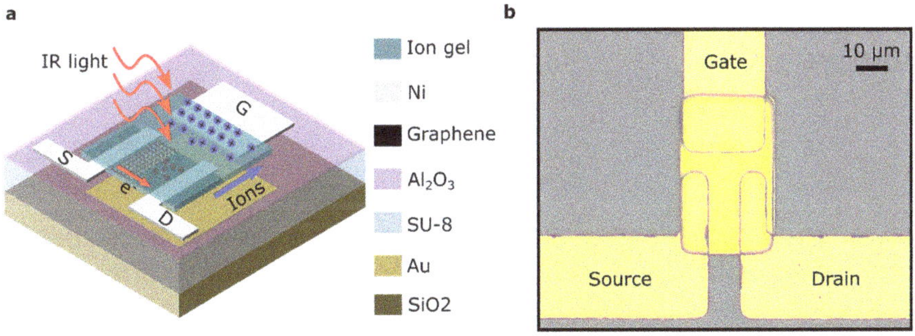

Fig. 2.3 Graphene-based LWIR detector architecture. (**a**) Schematic illustration and (**b**) optical microscope image of the detector

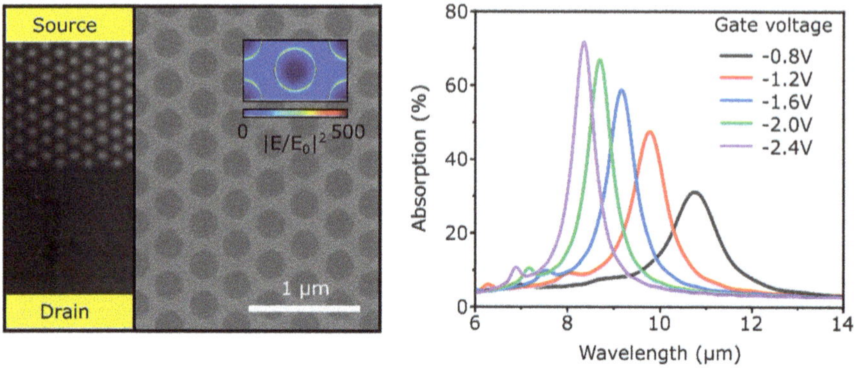

Fig. 2.4 Graphene device image and absorption. (**a**) SEM image of the asymmetrically patterned graphene and the corresponding zoomed-in view shows the hexagonal hole array pattern with diameter $D = 400$ nm and period $P = 600$ nm. The inset depicts a simulated electric field profile at resonance of the patterned graphene. (**b**) Simulated absorption spectra of the cavity-coupled graphene patterned with holes in a hexagonal array, shown as a function of the applied gate voltage

temperature is hardly affected. The Seebeck effect rises from the asymmetry of carrier temperature between the patterned and the unpatterned half of the graphene [29]. The diffusion of the hot carriers from the hot end to the cold end creates a photo-thermoelectric voltage across the source-drain electrodes [30, 31].

For efficient detection, the hot carriers need to be collected before transferring their energy to the lattice. Therefore, the channel length of the NPG between the source and the drain is designed to be 10 μm, which is shorter than the hot carriers' diffusion length which is estimated to be ~12 μm [12]. Figure 2.4b shows the dynamic tuning of the absorption spectra of the patterned graphene within the 6 μm to 14 μm as a function of the applied gate voltage. With an increase in gate voltage, a noticeable blue shift and enhanced absorption are observed. By increasing the gate voltage, the Fermi level of graphene is elevated, resulting in a higher carrier concentration and longer carrier lifetime due to the increase in density of states. This leads to increased conductivity, a stronger plasmonic response, and enhanced peak absorption. Conversely, decreasing the gate voltage reduces the response. The shift in localized surface plasmon resonance (LSPR) frequency with Fermi energy can be expressed by:

$$\omega_{\mathrm{LSPR}} \propto \sqrt{\frac{CE_F}{d}} \tag{2.1}$$

where ω_{LSPR} is the localized surface plasmon resonance frequency, d is the edge-to-edge distance of the adjacent anti-disks, E_F is the fermi energy of the graphene, and C is the eigenvalue of the total electric potential equation. Based on this equation, the elevated Fermi energy caused by a high gate voltage leads to an increased LSPR frequency, which corresponding to a blueshift of the absorption peak [18, 32]. It is worth noting that the absorption peaks exhibit a smaller full width at half maximum (FWHM) with a higher gate voltage, due to the longer carrier lifetime at high Fermi energy. The maximum absorption of 70% occurs at a wavelength of 8.3 μm for a gate voltage of −2.4 V, corresponding to the Fermi level of $E_F = 1$ eV. The overlay of the electrical field in Fig. 2.4a illustrates the finite difference time domain (FDTD)-predicted electrical field on the surface of the graphene at Fermi level $E_F = 1$ eV. The enhancement of the electrical field at the edge of the circular nano-hole patterns can be clearly observed.

The modulation of carrier concentration is assessed by measuring the resistance of the graphene as a function of the gate voltage. Figure 2.5 shows the gate-dependent resistance change. The n-type and p-type doping regions are highlighted with a red/blue background. The insets of Fig. 2.5 schematically show the band diagrams of graphene corresponding to both the doping types. In the case of p-type doping, the Fermi level is shifted toward the valence band, while in the case of n-type doping, it is shifted toward the conduction band. All the measurements are performed at room temperature. Here, the gate voltage V_g is expressed as $V_g = V_{gexp} - V_{CNP}$, where V_{gexp} is the applied gate voltage and V_{CNP} is the specific gate voltage where the charge-neutral point (CNP) of the graphene is found. With pristine graphene, CNP is expected to occur at $V_{CNP} = 0$ as the graphene's unique Dirac cone band structure leads to a zero-carrier concentration without gating. However, slight doping is common in CVD-grown and wet-transferred graphene due to defects and impurities introduced by various experimental conditions [33–35]. Thus, V_{CNP} is often found to be non-zero. Therefore, V_g instead of V_{gexp} is used in the figures to better present n-type and p-type doping. For $V_g < 0$, graphene undergoes

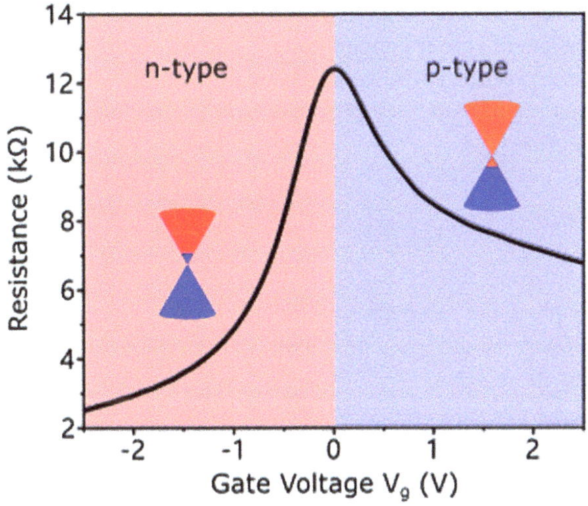

Fig. 2.5 Electrostatic doping of graphene. Change of carrier concentration measured as resistance change of the asymmetrically patterned graphene as a function of the gate voltage (V_g) at a constant source-drain voltage (V_{sd}) of 0.9 V

n-type doping, while $V_g = 0$ corresponds to a charge-neutral point, and $V_g > 0$ leads to p-type doping in graphene. Figure 2.5 shows a fourfold resistance change between $V_g = 0$ and $V_g = -2.4$ V in the n-type doping region, while a two-time resistance change is found with the p-type doping region. This asymmetry can be attributed to the CVD grow method, effect of substrates, and the impurities introduced during device fabrication, leading to a difference in electron and hole mobilities [36–38].

2.3.3 Characterizations of the Detector Performance

First, the photo response from the detector is characterized by illuminating it with a broadband blackbody light source and incident power of $P_{in} = 153$ nW. Figure 2.6a illustrates the changes in the thermoelectric voltage (V_{TE}) of the graphene detector with varying gate voltage under both blackbody light source (red) and dark (blue) conditions. During the measurements, the graphene channel is driven with a constant bias voltage of $V_{sd} = 0.9$ V between the source and the drain. The photo-thermoelectric voltage (V_{PTE}) is further extracted by subtracting the recorded V_{TE} values obtained in light and dark conditions. The experimentally measured V_{PTE} (black) along with the simulated V_{PTE} (red dashed) as a function of the gate voltage is plotted in Fig. 2.6b. The experimental result shows good agreement with the simulation. More details about the calculation of V_{TE} and V_{PTE} are described in the method section. Although the highest absorption occurs at the largest gate voltage $V_g = -2.4$ V (as can be seen Fig. 2.4b), the largest V_{PTE} is observed at a lower gate voltage $V_g = -1.4$ V. This can be explained by the origin of the photo-thermoelectric voltage from the Seebeck effect. The photo-thermoelectric voltage (V_{PTE}) can be mathematically represented as the integral of the Seebeck coefficient ($S(x)$) multiplied by the gradient of carrier temperature ($T_{carrier}$) which depends on the absorption with respect to the position (x) of the graphene film, ranging from the position of the electrical contact on the patterned graphene side (X_p) to the position on the unpatterned graphene side (X_u):

$$V_{PTE} = \int_{X_p}^{X_u} S(x) \frac{\partial T_{carrier}(x)}{\partial x} dx \tag{2.2}$$

Figure 2.6c plots the Seebeck coefficient S of the graphene film, which is extracted from the resistance-vs-gate voltage measurement [12, 25] in Fig. 2.5 with the Mott relation as:

$$S = \frac{\pi^2 k_B{}^2 T}{3e} \frac{\partial \ln \sigma}{\partial E_F} \tag{2.3}$$

where σ, k_B, and e are electrical conductivity, Boltzman constant, and electron charge, respectively. Here, E_F is related to the applied gate voltage by $E_F{=}\hbar v_F$

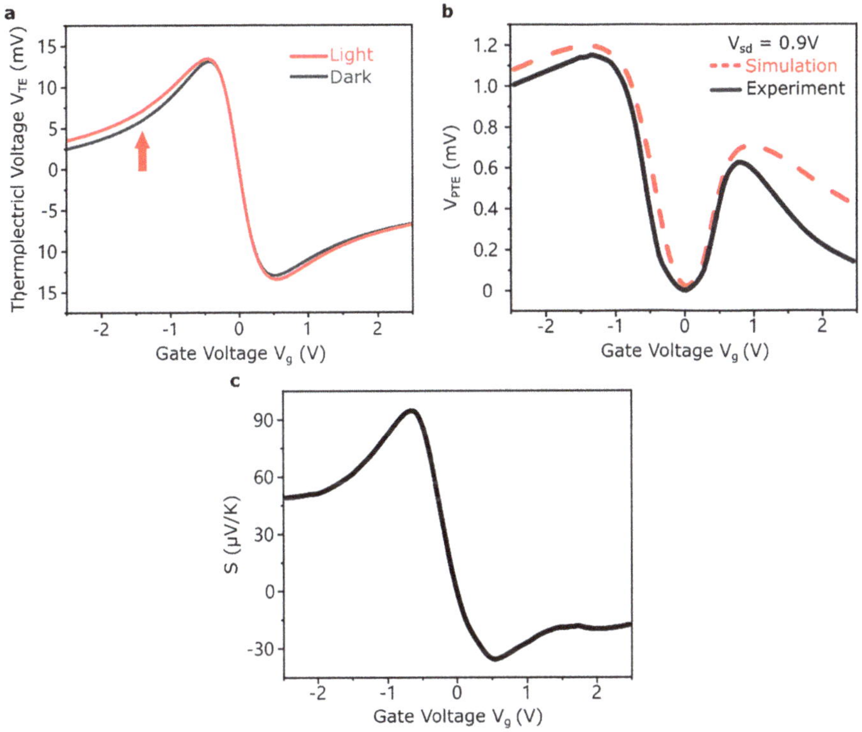

Fig. 2.6 Thermoelectric effect. (**a**) Thermoelectric voltage (V_{TE}) measurements corresponding to illumination with a blackbody light source and dark conditions plotted as a function of the gate voltage at the same V_{sd} of 0.9 V. (**b**) Comparison between simulated and experimental photo-induced thermoelectric voltage (V_{PTE}) as a function of gate voltage. (**c**) Gate voltage dependence of the Seebeck coefficient, deduced from the resistance vs. V_g measurement

$(\pi C V_g/e)^{1/2}$, where $v_F \approx 10^6$ ms^{-1} is the Fermi velocity of the charge carriers in graphene and C is the capacitance of the ion-gel gating medium. It can be observed that the corresponding V_g with the highest Seebeck coefficient is reached at $V_g = -0.75$ V. The combined impact of $T_{carrier}$ (x) and S contributes to the photo-response with respect to the gate voltage as can be seen in Fig. 2.6b.

The experimental findings are compared with the finite element simulations using COMSOL multiphysics, which considers both the carrier temperature and Seebeck coefficient. In the simulation, a constant source-drain voltage of $V_{sd} = 0.9$ V and an incident LWIR radiation power of 153 nW are maintained. The top image in Fig. 2.7a illustrates the distribution map of the carrier temperature across the half-patterned graphene channel, revealing a temperature difference of $\Delta T = 9.78$ K compared to the coolest region in the unpatterned graphene for a gate voltage of $V_g = -1.34$ V. The corresponding potential distribution map is displayed in the bottom image of Fig. 2.7a, demonstrating that the carrier temperature difference results in a photo-thermoelectric voltage of 1.2 mV. Furthermore, Fig. 2.7b presents line

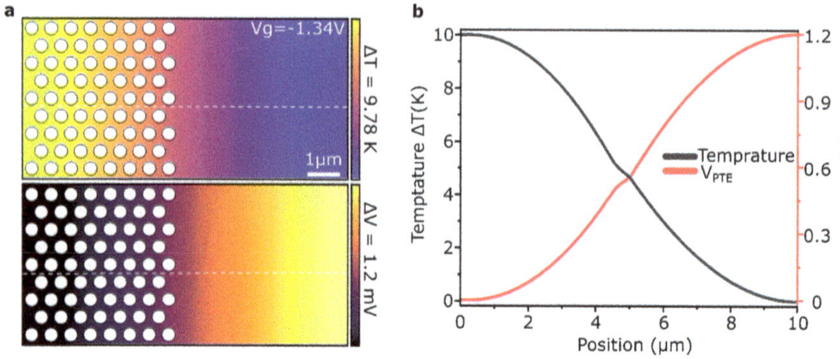

Fig. 2.7 V_{PTE} and temperature distribution. (**a**) Distribution maps of carrier temperature (top) and corresponding photo-thermoelectric potential (bottom) across the asymmetrically patterned graphene channel at $V_g = -1.34$ V. (**b**) Line-cut plots presenting the variation in carrier temperature and photo-thermoelectric voltage along the graphene channel

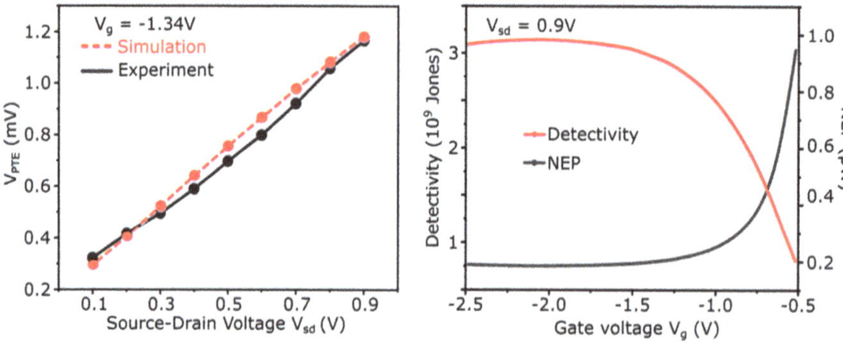

Fig. 2.8 Photo-thermoelectric voltage and detectivity. (**a**) Change of photo-thermoelectric voltage (V_{PTE}) as a function of source-drain voltage (V_{sd}) at a constant $V_g = -1.34$ V. (**b**) The effect of gate voltage on the detectivity (D^*) and noise equivalent power (NEP) at an optimum $V_{sd} = 0.9$ V

plots depicting the variation of carrier temperature and photo-thermoelectric voltage along the graphene channel in the temperature and potential maps, respectively. A clear trend is evident: as we move along from the patterned half to the unpatterned half, the carrier temperature decreases, while the photo-thermoelectric voltage increases. To investigate the correlation between the photo-thermoelectric voltage (V_{PTE}) and the source-drain voltage (V_{sd}), we keep the gate voltage fixed at $V_g = -1.34$ V while varying V_{sd} from 0 to 0.9 V. Both the experimental and simulation results, shown in Fig. 2.8a, indicate a clear linear relationship between V_{PTE} and V_{sd}. This behavior can be attributed to the increased diffusion of hot carriers facilitated by higher bias voltages. Although a higher V_{sd} leads to higher V_{PTE}, it is observed in the experiment that a V_{sd} of more than 1 V affects the stability of the graphene detector under constant operation. Thus, a source-drain voltage of 0.9 V is used in experiments for the optimal performance of the detector.

In order to evaluate the LWIR detector's performance, we conducted measurements on its dynamic tunability of detectivity, denoted as D^*. Figure 2.8b plots D^* and the noise equivalent power (NEP) with respect to the gate voltage. The D^* is defined by the expression:

$$D^* = \sqrt{A} / \mathrm{NEP} \tag{2.4}$$

where, A is the active area of the detector and NEP is the noise equivalent power. The NEP can be determined by:

$$\mathrm{NEP} = S_n / \mathcal{R} \tag{2.5}$$

where S_n is the noise spectral density measured at 1 Hz. The responsivity R can be determined as follows:

$$\mathcal{R} = V_{\mathrm{PTE}} / \mathrm{P}_{inc} \tag{2.6}$$

where Pinc is the incident power on the active region of the device. The highest detectivity of 3.15×10^9 Jones is measured at a gate voltage of −2.0 V. It is worth noting that the detectivity of the graphene detector exhibits a plateau over a wide range of gate voltage V_g between −2.5 and −1.5 V, which defines the optimum range of operation.

Next, we examined the dynamic spectral tunability of the monolayer graphene-based room-temperature LWIR detector. The correlation between the peak absorption wavelengths of the detector and the applied gate voltage is simulated using finite-difference time-domain (FDTD) simulation.

The contour plot in Fig. 2.9a shows the evolution of the absorption wavelength as a function of the gate voltage. Furthermore, in Fig. 2.9b, the experimentally measured absorption spectra for different gate voltages are compared with the FDTD simulation results. The measured absorption spectra are represented by solid lines, while the simulated results corresponding to each gate voltage are shown as dashed lines of same colors. It is evident from the experimental and the simulation data that a broad tuning range of 2.7 μm for the center wavelength is achieved, spanning from approximately 8.5 to 11.2 μm.

Figure 2.10a illustrates the peak detectivities at the selected gate voltages plotted against the corresponding peak absorption wavelengths. This further demonstrates the dynamic spectral tunability of the graphene-based room-temperature LWIR detector across the atmospheric transparent window. The time response of the proposed detector is measured by recording the rise time τ_{rise} and fall time τ_{fall} during the switching between dark-to-light and light-to-dark conditions, respectively. τ_{rise} and τ_{fall} are calculated by the time duration between 10% and 90% of changes between the steady-state voltages. The total time response of the detector is further expressed as $\tau_{\mathrm{total}} = \tau_{\mathrm{rise}} + \tau_{\mathrm{fall}}$.

Figure 2.10b plots the experimental data of the time response measurement, where we record a $\tau_{\mathrm{rise}} = 74.6$ ns and $\tau_{\mathrm{fall}} = 69.5$ ns indicating a response time of

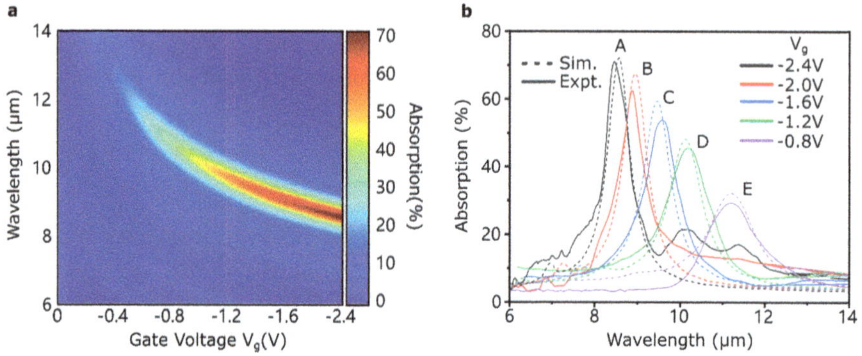

Fig. 2.9 Spectrally tunable LWIR detection. (**a**) Simulated contour plot depicting the evolution of the absorption peak of the hexagonal hole array patterned graphene as a function of gate voltage. (**b**) Comparison of the experimentally measured (solid) and simulated (dashed) absorption spectra of the patterned graphene at certain gate voltages for a constant $V_{sd} = 0.9$ V

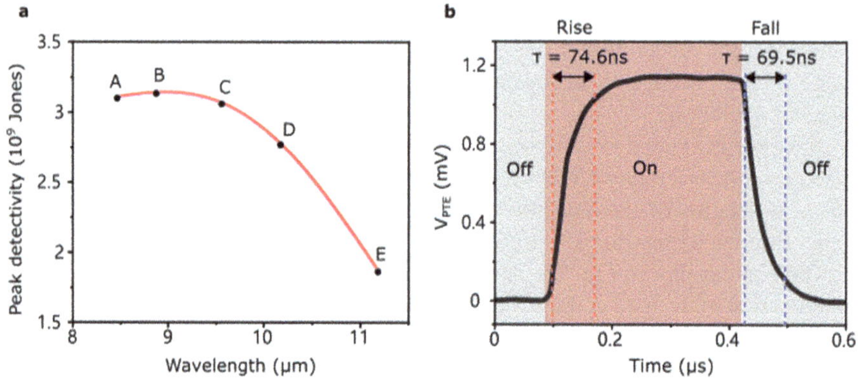

Fig. 2.10 Spectrally tunable LWIR detection. (**a**) The measured peak detectivity vs. peak absorption wavelength. (**b**) Time response measurement of the graphene detector showing an ultra-fast response time of $\tau \sim 144$ ns

$\tau_{total} = 144$ ns. It should be noted that the response time of our detector is only limited by the experimental setup. The response time can be further reduced as the Dirac plasmon relaxation lifetime is less than 1 ps [39]. This response time is three orders of magnitude faster compared to the cryogenically cooled detectors which are typically limited to the micro-second response time. High sensitivity ($D^* > 10^9$), fast response time ($\tau < 150$ ns), and dynamic spectral tunability at room temperature make this proposed detector a new class of uncooled LWIR detector which was missing from the present technologies.

2.4 Scatter-Type Scanning Nearfield Optical Microscopy (s-SNOM)

Scattering-type scanning nearfield optical microscopy, commonly abbreviated as s-SNOM, is an advanced imaging technique used in nanoscale and surface science research. The configuration of the s-SNOM is schematically shown in Fig. 2.11. It consists of an AFM system with external optical component to illuminate the tip with external laser and collect the back scattered light. S-SNOM is a powerful analytical method that allows scientists to investigate and visualize nanoscale structures and interactions at the molecular and atomic levels. s-SNOM employs the principles of nearfield microscopy and spectroscopy to achieve high-resolution imaging and spectroscopic analysis, enabling researchers to study materials and devices with unprecedented detail.

Key features of s-SNOM include nearfield imaging, high spatial resolution (surpassing the diffraction limit), and spectroscopic capabilities. For nearfield imaging, s-SNOM operates in the nearfield region, which is very close to the sample surface, typically within a few nanometers. Thus, the scattered signal collected contains nearfield information and differs from a far-field microscopy. And by detecting the scattered light with an AFM tip of radius ~10 s nm, s-SNOM can achieve spatial resolution well below the diffraction limit of light. This enables researchers to visualize and analyze nanoscale features, including surface roughness, chemical composition, and structural details. In addition to imaging, s-SNOM can also perform spectroscopy. It can measure vibrational, electronic, and other optical properties of materials at the nanoscale, providing insights into chemical composition and material behavior.

Applications of s-SNOM span a wide range of scientific disciplines, including materials science, biology, and condensed matter physics. Researchers use s-SNOM to study surface plasmon resonance, investigate nanoscale structures, and gain a deeper understanding of the interactions and properties of various materials. This technique has become invaluable for advancing our understanding of nanoscale phenomena and for the development of cutting-edge technologies in fields like nanotechnology, photonics, and materials research.

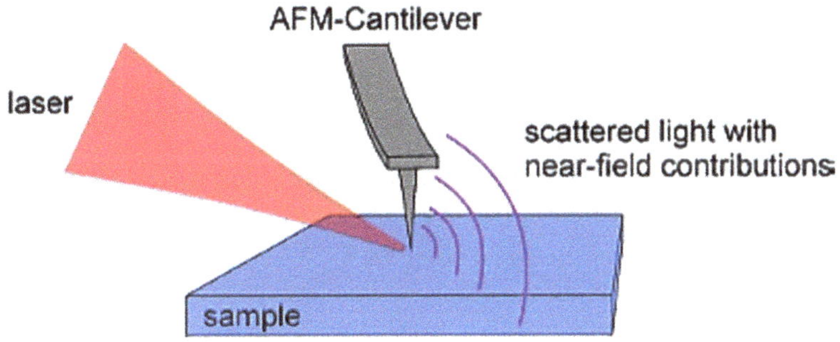

Fig. 2.11 Configuration of the s-SNOM

2.4.1 Imaging of the Plasmon in Decorated Monolayer Graphene

Propagating surface plasmons, often referred to as surface plasmon polaritons (SPPs), are collective electron oscillations or waves that propagate along the interface between a metal and a dielectric or insulating material. These waves result from the interaction between incident light and the free electrons at the metal-dielectric boundary. Propagating surface plasmons have garnered significant interest in the fields of photonics, plasmonic, and nanophotonic due to their unique properties and potential applications. Understanding and controlling propagating surface plasmons have led to groundbreaking advancements in the field of photonics and have opened new possibilities for developing compact and highly sensitive optical devices. The ability to manipulate light at the nanoscale using SPPs has implications in various fields, including telecommunications, biosensing, imaging, and data storage, making it a topic of significant research and innovation.

Here, we demonstrate some preliminary results of imaging the SPPs in plasmonic structure–decorated monolayer graphene with s-SNOM. The structures are a series of holes with different diameters ranging from 160 to 300 nm. With an incident light illumination on the interface of graphene and metal AFM tip, the tip-enhanced field excites the SPPs in graphene. With the plasmonic structure decoration, the excited plasmons will interference with SPPs bounced back from the plasmonic structure decoration in graphene, forming interface patterns as shown in Fig. 2.12. Both the amplitude and phase of the scattered field are recorded by s-SNOM. The amplitude of the scattered field mainly contains information about the reflection properties of the surface. The field amplitude is strongest where the nearfield is most intense, which is the edge of the ring structure as shown in Fig. 2.12a. The scattered phase, on the other hand, reflects the absorption-related properties of the probed interface and, thus, often possesses greater contrast compared to the amplitude of the signal near peak of absorption. In Fig. 2.12b, the phase of the scattered nearfield clearly showed characteristics of interfering SPPs.

2.4.2 Nearfield Imaging of the Hot Carrier Decay in Graphene

Here we demonstrate the imaging of hot carrier decay in the proposed LWIR graphene detector. To achieve this, we need an additional light source apart from the one which illuminates at the tip-graphene interface as shown in Fig. 2.13. The purpose of the additional QCL source is to excite the patterned graphene with hot carriers and study their drift.

To eliminate the unwanted signal from the QCL source, we intentionally misalign the QCL source with the tip-graphene interface, such that the scattered light from the QCL source is minimal. It is worth noting that the alignment of the incident light to the AFM tip is critical to get any scattered signal to the detector in

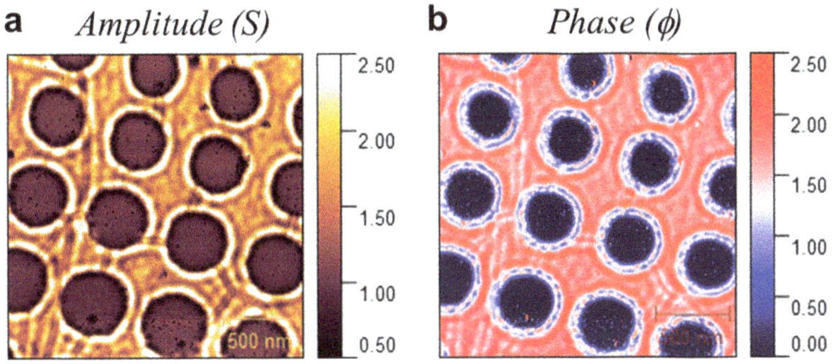

Fig. 2.12 SPPs in graphene. (**a**) Scattered electromagnetic field amplitude and (**b**) scattered field phase

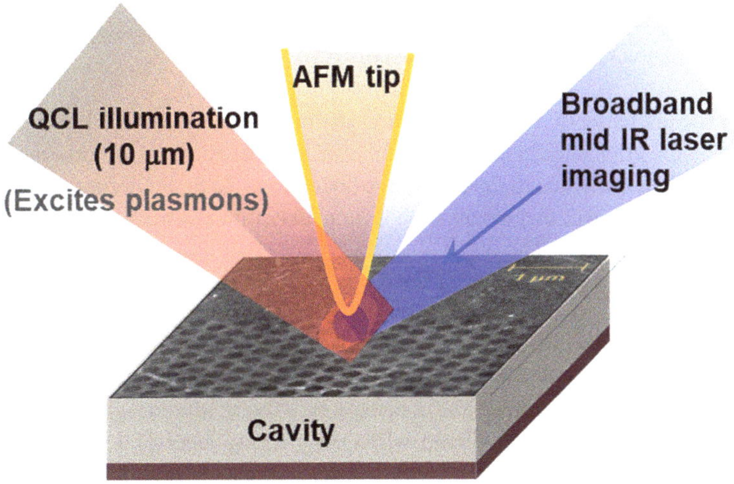

Fig. 2.13 Additional illumination for hot carrier generation. Additional light source is for the excitation of hot carriers in graphene

s-SNOM, a misalignment of 1 ~ 2 μm drastically decreases the signal to about >20 times or more. With this, we misalign the QCL source intentionally while making sure the beam spot covers the whole graphene channel.

With the above, we performed s-SNOM measurement with an additional excitation of QCL laser at 10 μm on half-patterned graphene as previously introduced. Figure 2.14a shows the s-SNOM imaging with different QCL laser power from 0 to 4.7 mW. With all experiment, the effect of the scattered QCL signal collected by the detector in s-SNOM is carefully examined with switching on and off the QCL illumination. While similar signal intensity is observed with the switching on/off the QCL laser, and the standby signal is much less compared to while taking a measurement, it validates the actual measurement from the possible noise contribution with the QCL laser.

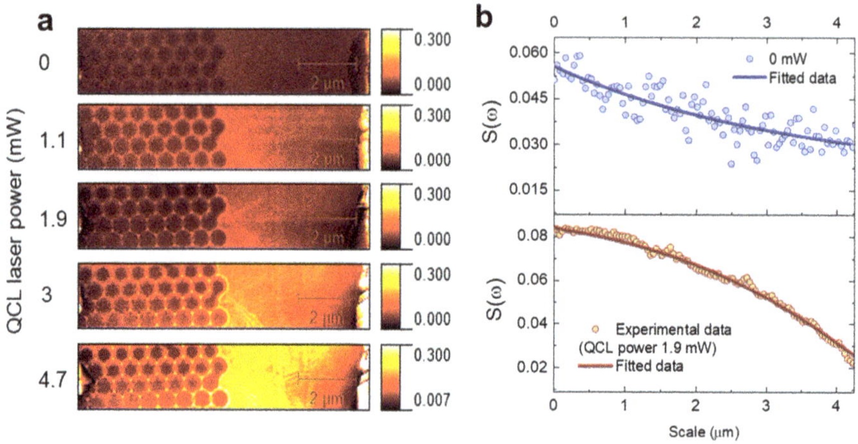

Fig. 2.14 Hot carrier drift in graphene. (**a**) Hot carrier drifting in half-patterned graphene with different laser power. (**b**) Line fit of the carrier drift

With the increase of the QCL power, an increased number of hot carriers is observed in the graphene channel. In the meantime, local hot spots are observed in the patterned region likely due to the drift of the hot carriers. In this experiment, no source drain current is applied, this means both source and drain electrodes can act as a collector of the hot carriers. The result aligns with this assumption as the scattered field intensity is strongest at the center of the channel which has an equal distance to both electrodes. Because of the nano-patterns in the patterned region, it is not trivial to observe the hot carrier drifting, here we investigate the unpatented region.

As shown in Fig. 2.14b, the line plot of the unpatterned channel at 2 QCL laser power 0 and 1.9 mW is plotted. The distribution of the scattered signal with a 0 mW laser power is due to the broadband s-SNOM imaging laser of 1 mW. While the QCL laser is a single wavelength laser and possesses much more power in the resonance region of graphene, we plotted the result with a 1.9 mW power as shown in Fig. 2.14b.

2.5 Discussion

In conclusion, we have successfully demonstrated a room-temperature LWIR detector with dynamic spectral tunability using a cavity-coupled nano-patterned monolayer graphene. Our work is significant as it harnesses the unique properties of graphene, such as high carrier mobility, fast carrier relaxation time, and dynamic tunability of its optical properties through electrostatic gating to enable efficient long-wavelength infrared (LWIR) detection. By exciting Dirac plasmons on a hexagonal array of circular holes coupled to a Fabry-Perot cavity, the challenge of graphene's low absorption is successfully overcome, resulting in an approximately 30 times enhancement in absorption. The transparent nature of the ion-gel used in our design allows direct access of LWIR radiation to the graphene layer, enhancing its sensitivity. Moreover, by adjusting the gate voltage, the absorption of the nano-patterned graphene can be dynamically tuned. The graphene-based LWIR detector showcased exceptional performance, with a high detectivity of 3.15×10^9 Jones and a fast response time of 144 ns, surpassing the capabilities of conventional uncooled microbolometers by several orders of magnitude. This breakthrough paves the way for low-cost, sensitive, and fast infrared detection at room temperature. Importantly, our detector provides the advantage of dynamically tuning the spectral response by selectively targeting operational wavelengths within the long-wavelength infrared (LWIR) range of 8–14 μm, thereby addressing the limitations of both traditional cryogenically cooled LWIR photodetectors and bolometers that lack spectral tunability. The versatility of our approach opens exciting possibilities for a wide range of applications, including thermal imaging, night vision, and material identification. By exploiting the dynamic spectral tunability and superior performance of our graphene-based detector, these applications can benefit from improved sensitivity and fast time response. Overall, our work represents a significant step forward in the development of advanced room-temperature LWIR detection technology with broad implications in various fields.

In addition, to study the physics behind the detector, we have employed s-SNOM to image the SPPs in plasmonic structure–decorated graphene as well as hot carrier generation/drifting in half-patterned graphene.

References

1. Liu, J., Xia, F., Xiao, D., García de Abajo, F. J., & Sun, D. (2020). Semimetals for high-performance photodetection. *Nature Materials, 19*, 830–837. https://doi.org/10.1038/s41563-020-0715-7
2. Tan, C. L., & Mohseni, H. (2018). Emerging technologies for high performance infrared detectors. *Nanophotonics, 7*, 169–197. https://doi.org/10.1515/nanoph-2017-0061
3. Rogalski, A., Martyniuk, P., & Kopytko, M. (2016). Challenges of small-pixel infrared detectors: a review. *Reports on Progress in Physics, 79*, 046501. https://doi.org/10.1088/0034-4885/79/4/046501
4. Russell, T. A., McMackin, L., Bridge, B. & Baraniuk, R. in *Compressive Sensing*. 55–67 (SPIE).
5. Talghader, J. J., Gawarikar, A. S., & Shea, R. P. (2012). Spectral selectivity in infrared thermal detection. *Light: Science & Applications, 1*, e24–e24. https://doi.org/10.1038/lsa.2012.24
6. Lei, W., Antoszewski, J., & Faraone, L. (2015). Progress, challenges, and opportunities for HgCdTe infrared materials and detectors. *Applied Physics Reviews, 2*. https://doi.org/10.1063/1.4936577
7. Rogalski, A. (2005). HgCdTe infrared detector material: history, status and outlook. *Reports on Progress in Physics, 68*, 2267. https://doi.org/10.1088/0034-4885/68/10/R01
8. Ferguson, B., & Zhang, X.-C. (2002). Materials for terahertz science and technology. *Nature Materials, 1*, 26–33. https://doi.org/10.1038/nmat708
9. Richards, P. L. (1994). Bolometers for infrared and millimeter waves. *Journal of Applied Physics, 76*, 1–24. https://doi.org/10.1063/1.357128

10. Zhang, H., et al. (2015). MWIR/LWIR filter based on Liquid–Crystal Fabry–Perot structure for tunable spectral imaging detection. *Infrared Physics & Technology, 69*, 68–73.
11. Neumann, N., Ebermann, M., Kurth, S., & Hiller, K. (2008). Tunable infrared detector with integrated micromachined Fabry-Perot filter. *Journal of Micro/Nanolithography, MEMS and MOEMS, 7*, 021004-021004-021009.
12. Safaei, A., Chandra, S., Shabbir, M. W., Leuenberger, M. N., & Chanda, D. (2019). Dirac plasmon-assisted asymmetric hot carrier generation for room-temperature infrared detection. *Nature Communications, 10*, 3498.
13. Guo, Q., et al. (2018). Efficient electrical detection of mid-infrared graphene plasmons at room temperature. *Nature Materials, 17*, 986–992.
14. Xia, F., Farmer, D. B., Lin, Y.-m., & Avouris, P. (2010). Graphene field-effect transistors with high on/off current ratio and large transport band gap at room temperature. *Nano Letters, 10*, 715–718.
15. Tsang, K. H., D., et al. (2019). Chemically functionalised graphene FET biosensor for the label-free sensing of exosomes. *Scientific Reports, 9*, 13946.
16. Ergoktas, M. S., et al. (2021). Multispectral graphene-based electro-optical surfaces with reversible tunability from visible to microwave wavelengths. *Nature Photonics, 15*, 493–498.
17. Safaei, A., et al. (2017). Dynamically tunable extraordinary light absorption in monolayer graphene. *Physical Review B, 96*, 165431. https://doi.org/10.1103/PhysRevB.96.165431
18. Safaei, A., Chandra, S., Leuenberger, M. N., & Chanda, D. (2018). Wide angle dynamically tunable enhanced infrared absorption on large-area nanopatterned graphene. *ACS Nano, 13*, 421–428.
19. Cakmakyapan, S., Lu, P. K., Navabi, A., & Jarrahi, M. (2018). Gold-patched graphene nanostripes for high-responsivity and ultrafast photodetection from the visible to infrared regime. *Light: Science & Applications, 7*, 20.
20. Kim, B. J., et al. (2010). High-performance flexible graphene field effect transistors with ion gel gate dielectrics. *Nano Letters, 10*, 3464–3466.
21. Wang, D., et al. (2021). Recent advanced applications of ion-gel in ionic-gated transistor. *Npj Flexible Electronics, 5*, 13.
22. Oh, G., et al. (2015). Graphene/pentacene barristor with ion-gel gate dielectric: Flexible ambipolar transistor with high mobility and on/off ratio. *ACS Nano, 9*, 7515–7522.
23. Lee, S. W., et al. (2015). Thin ion-gel dielectric layer to enhance the stability of polymer transistors. *Science of Advanced Materials, 7*, 874–880.
24. Xu, X., Gabor, N. M., Alden, J. S., Van Der Zande, A. M., & McEuen, P. L. (2010). Photo-thermoelectric effect at a graphene interface junction. *Nano Letters, 10*, 562–566.
25. Shautsova, V., et al. (2018). Plasmon induced thermoelectric effect in graphene. *Nature Communications, 9*, 5190.
26. Gabor, N. M., et al. (2011). Hot carrier–assisted intrinsic photoresponse in graphene. *Science, 334*, 648–652.
27. Yan, H., et al. (2013). Damping pathways of mid-infrared plasmons in graphene nanostructures. *Nature Photonics, 7*, 394–399.
28. Buljan, H., Jablan, M., & Soljačić, M. (2013). Damping of plasmons in graphene. *Nature Photonics, 7*, 346–348.
29. Harzheim, A., et al. (2018). Geometrically enhanced thermoelectric effects in graphene nanoconstrictions. *Nano Letters, 18*, 7719–7725.
30. Sierra, J. F., Neumann, I., Costache, M. V., & Valenzuela, S. O. (2015). Hot-carrier Seebeck effect: diffusion and remote detection of hot carriers in graphene. *Nano Letters, 15*, 4000–4005.
31. Ruzicka, B. A., et al. (2010). Hot carrier diffusion in graphene. *Physical Review B, 82*, 195414.
32. Fang, Z., et al. (2014). Active tunable absorption enhancement with graphene nanodisk arrays. *Nano Letters, 14*, 299–304.
33. Wang, Q. H., et al. (2012). Understanding and controlling the substrate effect on graphene electron-transfer chemistry via reactivity imprint lithography. *Nature Chemistry, 4*, 724–732.

34. Gosling, J. H., et al. (2021). Universal mobility characteristics of graphene originating from charge scattering by ionised impurities. *Communications Physics, 4*, 30.
35. Chen, J.-H., et al. (2008). Charged-impurity scattering in graphene. *Nature Physics, 4*, 377–381.
36. Li, M.-Y., et al. (2013). Charged impurity-induced scatterings in chemical vapor deposited graphene. *Journal of Applied Physics, 114*.
37. Van Veldhoven, Z. A., Alexander-Webber, J. A., Sagade, A. A., Braeuninger-Weimer, P., & Hofmann, S. (2016). Electronic properties of CVD graphene: The role of grain boundaries, atmospheric doping, and encapsulation by ALD. *Physica Status Solidi (B), 253*, 2321–2325.
38. Negishi, R., Ohno, Y., Maehashi, K., Matsumoto, K., & Kobayashi, Y. (2012). Carrier transport properties of the field effect transistors with graphene channel prepared by chemical vapor deposition. *Japanese Journal of Applied Physics, 51*, 06FD03.
39. Vincent, J. D., Hodges, S., Vampola, J., Stegall, M., & Pierce, G. (2015). *Fundamentals of infrared and visible detector operation and testing*. Wiley.

Chapter 3
Frequency Modulation–Based Infrared Detection

Abstract All existing LWIR detectors, whether they are of the cooled or uncooled variety, currently utilize amplitude modulation (AM) for the detection of changes in photocurrent, voltage, or resistance in response to light exposure. These AM-based detection methods are inherently susceptible to various forms of AM noise, including Johnson noise, flicker noise, and shot noise (Vincent, et al. Fundamentals of infrared and visible detector operation and testing. Wiley2015; Yadav, et al. Sensors Actuators A: Physical 342:113611. https://doi.org/10.1016/j.sna.2022.113611, 2022; Donati, Photodetectors: devices, circuits and applications. Wiley, 2021). Therefore, there is a critical demand for an uncooled LWIR detector that can provide superior detectivity and an improved signal-to-noise response.

Here, we introduce an LWIR detection approach based on frequency modulation (FM), utilizing an oscillating circuit integrated with a phase-change material. While traditional microbolometers operate by sensing changes in resistance due to small perturbations in material properties with LWIR light illumination, they often exhibit a positive resistance characteristic. Our proposed scheme functions through FM of the circuit. This modulation is achieved by harnessing the negative differential resistance (NDR) property change of the phase-change material with incident light, which undergoes dramatic phase transitions between insulating and metallic states.

This FM-based detection scheme inherently offers robustness against noise, especially AM noises. Our room-temperature LWIR detector demonstrates a noise equivalent power (NEP) of less than 3 pW·Hz$^{-1/2}$, a response time of approximately 2.96 ms, and a high detectivity (D*) on the order of 10^9. The performance of the FM based detection in this work can be further improved through appropriate industry--scale packaging with reduced capacitance.

T. Guo, *Low Energy Photon Detection*, Springer Theses,
https://doi.org/10.1007/978-3-031-71544-0_3

3.1 Insulating to Metallic Phase Transitions

3.1.1 *Phase-Change Material*

Lately, there has been a growing fascination with phase-change materials (PCMs) as a promising avenue for advancing adaptive materials that have the potential to revolutionize the realm of metamaterial-based applications. PCMs, as a category of materials, can undergo substantial changes in their material phase when exposed to external stimuli, such as conductive heating, optical activation, or the application of electric or magnetic fields. These transformations lead to alterations in their electrical and optical characteristics. Notably, PCMs exhibit varying levels of volatility, with some falling into the category of volatile PCMs, requiring continuous stimulation to maintain their altered state, while others are categorized as non-volatile PCMs, preserving their phase change even after the stimulus is removed until a specific level of stimulus is reapplied. This diversity in volatility is a significant consideration when dealing with PCMs, and researchers are actively exploring innovative ways to harness this distinctive property for the development of more dynamic and adjustable materials.

In this study, we employ volatile phase-change materials (PCMs) that undergo phase transitions in response to external stimuli, reverting to their original states when the stimulus is removed. Our specific focus lies in PCMs that exhibit alterations in their characteristics upon exposure to LWIR light. By detecting these changes in properties, we accomplish LWIR detection.

3.1.2 *Thin Film Vanadium Dioxide*

Unravelling the physical characteristics of strongly correlated systems has remained one of the most intricate challenges within the field of condensed matter physics, persistently raising contentious questions both in theory and in experiment. Among the materials of interest, the V–O system stands out as one of the most captivating due to its diverse array of compounds, including V_2O_3, VO_2, and V_2O_5, among others. Many of these compounds undergo a distinctive metal-to-insulator phase transition (MIT), resulting in abrupt alterations in their optical, electrical, thermal, and magnetic properties. This system not only holds considerable promise for practical applications but also offers scientists compelling and important materials for investigating the mechanisms behind phase transitions.

Vanadium dioxide (VO_2) is particularly intriguing within the V–O system because its transition temperature lies in proximity to room temperature. Despite ongoing debate over its phase-transition mechanism, there has been notable progress in modifying this transition in recent years [4]. The crystal structure of different VO_2 phases with the corresponding band structures can be seen in Fig. 3.1 [4].

The behavior of MIT in VO_2 was initially documented by Morin in 1959. The transition from a first-order monoclinic (M) insulator phase to a rutile (R) metallic

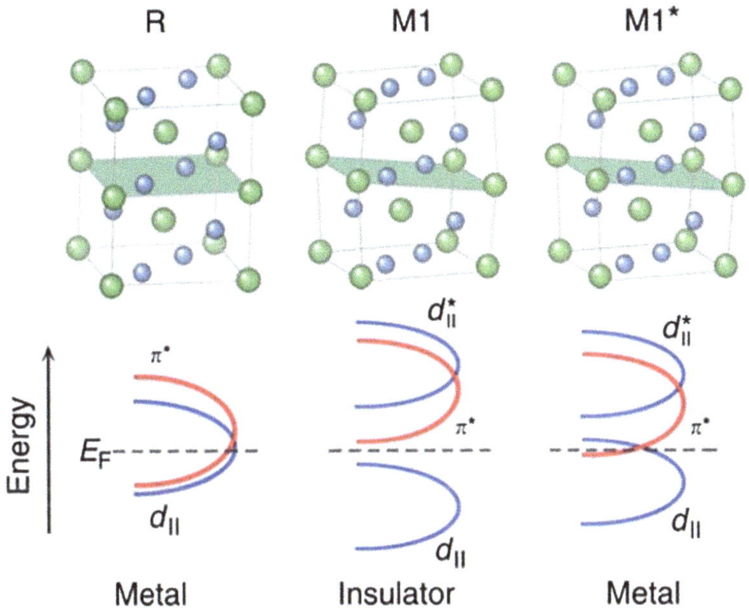

Fig. 3.1 Crystal structure of VO_2 and corresponding band structures [4]. The band structure of metal and insulator phase of VO_2

phase in VO_2 is accompanied by a significant and rapid transformation in electrical and optical properties, with potential applications in switching, sensing, and the control of infrared (IR) wavelengths. Recently, VO_2 has gained renewed attention as a conveniently accessible strongly correlated material, thanks to advances in ultrafast and microscopy techniques, ionic gating, and improved computational methods. Leveraging these new research tools, considerable efforts have been invested in various aspects, such as exploring different material structures (films, nanobelts, nanowires, etc.), refining fabrication techniques, uncovering phase-transition mechanisms, and modulating MIT behaviors. Several studies have presented different morphologies of VO_2 using diverse fabrication methods, resulting in enhanced performance. Furthermore, gaining a deep understanding of the mechanisms underlying phase transitions not only improves material performance and broadens its applications but also contributes to a more profound comprehension of the phase-transition process.

3.1.3 Mechanisms of Phase Transitions

In the past two decades, the exploration of the physical properties of phase-transition systems has presented one of the most formidable challenges in the realm of condensed matter physics. This field continues to grapple with contentious issues both

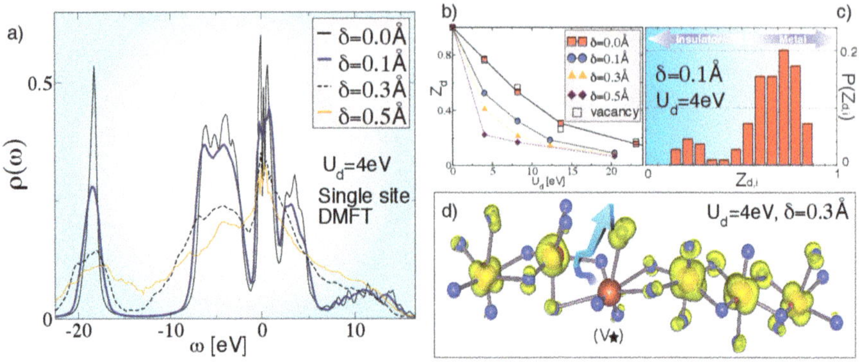

Fig. 3.2 Spectral function of VO_2 and its crystal structure [5]. (**a**) Spectral function ρ of paramagnetic VO_2 in the presence of Gaussian disorder δ. (**b**) Averaged quasiparticle weight Z_d. (**c**) Distribution of the local quasiparticle weight Zd,i for δ = 0.1 Å. (**d**) Isosurface of the real space representation of the Fermi density for disorder δ = 0.3 Å. The large (small) sphere denotes V (O) atoms along the rutile axis [5]

in theory and in experimentation. Within the context of the VO_2 system's transition, several simultaneous phenomena, including Mott correlation, Peierls distortion, and orbital splitting in the V–V chains, have complicated our comprehension of the underlying mechanism. This chapter aims to shed light on the focal points of mechanism research and the significance of understanding the phase-transition characteristics of VO_2 for subsequent modulation. A central point of controversy revolves around the theoretical interpretation of lattice-electron interactions and electron--electron correlations during the phase transition. The VO_2 in its rutile axis and the researcher's effort of discovering the phase transition mechanics are shown in Fig. 3.2 [5].

Back in the 1930s, Rudolf Peierls postulated that a 1D conductor comprises atoms with uniform spacing, each associated with a conduction electron. Due to the interaction between electrons and the lattice, this 1D system becomes unstable, resulting in lattice distortions at pairs of atoms at low temperatures. When the modulation wave vector q equals 2 times Fermi wave vector k, the Fermi surface aligns perfectly with the boundary of the Brillouin zone, causing complete "nesting." When the energy expended in creating the band gap is greater than the energy required for the lattice distortion, the overall electronic-lattice system's energy decreases, leading to the transition from a conductor to an insulator. Goodenough was the first to employ crystal field and molecular orbital theories to analyze the electronic structure of VO_2, attributing the MIT transformation to lattice distortion, thereby associating it with the Peierls transition. However, it's worth noting that the MIT in VO_2 is considerably more complex, and the Peierls mechanism falls short in completely addressing all the issues related to the phase transition. Essentially, the Peierls mechanism is a result derived from traditional band theory, which may have overlooked crucial factors in the phase transition or, in other words, is not entirely suitable for explaining the behavior of VO_2 materials.

Later, as observed by Mott in his 1949 publication on Ni-oxide, the origins of this phase-change behavior are linked to electron correlations and their strong connection to magnetism. The Mott transition's physical underpinning arises from the interplay between the Coulomb repulsion among electrons and their degree of localization, represented by the band width. When the carrier density becomes sufficiently high, the system's energy can be minimized by localizing the previously conducting electrons (reducing the band width), which leads to the formation of a band gap.

3.2 Self-Oscillation in Electrical Circuits

3.2.1 *Self-Oscillators and Negative Differential Resistance*

In electrical circuits, self-oscillators refer to a circuit element that does not require an AC voltage/current input to output an AC voltage/current. Such self-oscillators can be created by utilizing circuit elements that exhibit NDR coupled with a proper capacitor.

While a typical resistor exhibits a change of current induced by a change of voltage $\Delta V/\Delta I > 0$, an element with NDR exhibits $\Delta V/\Delta I < 0$. They are widely used in oscillators, memory devices, frequency multipliers, mixers, etc. When such circuits are biased within the NDR region, small disturbances or noise signals can initiate oscillations, and the negative resistance will sustain and amplify these oscillations until a stable output waveform is achieved.

In this work, we utilize a self-oscillator with an NDR element. With light illumination, the property of the NDR element changes correspondingly resulting in the oscillation frequency change to the self-oscillator. By recording this change in frequency, the LWIR detection with FM scheme is realized.

3.2.2 *Vanadium Dioxide–Based Self-Oscillators*

In this work, VO_2 is chosen as the phase-change material [6]. At room temperature, VO_2 assumes an insulating phase characterized by a monoclinic crystal structure and high resistance [7–9]. However, when exposed to higher temperatures, VO_2 undergoes a transition, shifting from this insulating phase to a metallic phase, commonly referred to as the insulator-to-metal transition (IMT). In this metallic phase, VO_2 displays significantly reduced resistance, orders of magnitude lower than in its insulating state [10, 11]. This transition is reversible, allowing VO_2 to return to its insulating phase when the temperature decreases, which is known as the metal-to-insulator transition (MIT) [12, 13]. Notably, this thermally induced phase transition in VO_2 is particularly robust when compared to other phase-change materials, and it exhibits a narrow thermal hysteresis range of 5–8 K [14, 15].

Within the phase transition range, VO_2 demonstrates a unique characteristic known as negative differential resistance (NDR) [16], where the differential resistance ($\Delta V/\Delta I$) becomes negative. We leverage this property by combining an NDR element with a capacitor, resulting in an electrical oscillator circuit that can independently sustain current/voltage oscillations, eliminating the need for an external inductive component [17–20]. In the case of VO_2, this leads to repeated oscillation between its insulating and metallic phases, with the IMT and MIT transitions following one another.

To improve the detector's optical absorption and sensitivity, we place a thin film of VO_2 on an optical cavity. Our frequency modulation (FM)-based LWIR detection method relies on monitoring changes in the oscillator circuit's frequency concerning incident power while electrically biasing the VO_2 film near the IMT transition edge, where its temperature coefficient of resistance (TCR) is high.

Prior research has successfully demonstrated the creation of self-oscillators with NDR using various phase-change materials, including metal oxides and 2D materials [21–24]. The detection scheme presented in this work is compatible with such materials and changing the film from VO_2 to these materials may invoke distinct properties to the LWIR detector.

3.3 Experimental Methods

3.3.1 Principle of Operation

The schematic design of the FM-based LWIR detector is shown in Fig. 3.3a. A thin PCM film is suspended with a Si_3N_4 support layer on top of an air spacer and an Au mirror. The Au mirror and air spacer together form an optical cavity. A part of the PCM/Si_3N_4 stack in the schematic is intentionally drawn with transparency to clearly show the presence of the optical cavity. The incident infrared radiation (indicated by the red arrow) is partially absorbed and partially transmitted through the thin film stack. The transmitted wave is then reflected by the back mirror. Depending on the cavity phase, a constructive interference between the reflected wave and incident wave intensifies the electromagnetic field at the suspended PCM/Si_3N_4 stack interface resulting in an enhancement in absorption [25]. Finally, chromium electrodes were fabricated on the sides of the PCM for electrical connections.

A scanning electron microscope (SEM) image of the detector is shown in Fig. 3.3b, where a suspended $8 \times 8\ \mu m^2$ PCM/Si_3N_4 stack is clearly visible. Details of the device fabrication are included in the Methods section. The fabricated detector pixel is connected to an electrical circuit as illustrated in Fig. 3.3c. At an appropriate biasing condition, electrical oscillations can be initiated owing to the NDR exhibited by the phase-change material in the active pixel material stack. The onset of oscillation under current control (CC) [24] mode is depicted in Fig. 3.3c. The two halves in the circuit diagram, highlighted by blue and red colors, represent two

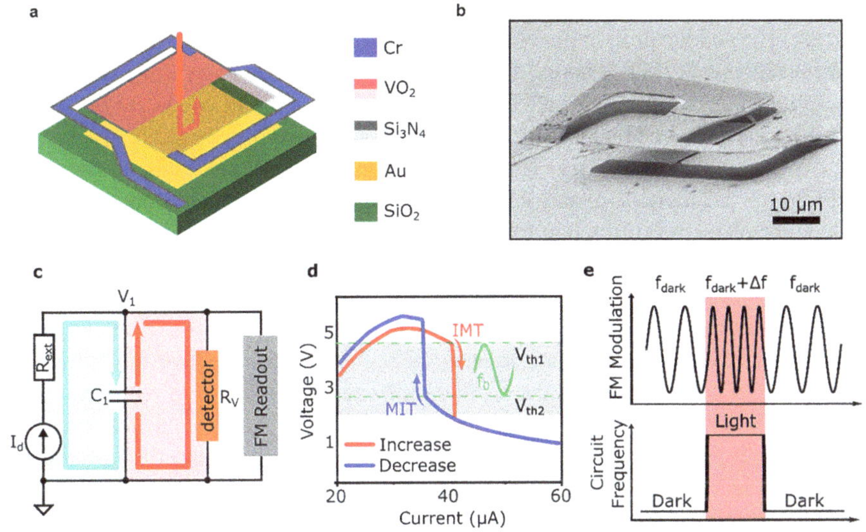

Fig. 3.3 FM-based infrared detection. (**a**) Schematic illustration and (**b**) corresponding scanning electron microscope (SEM) image of the device. (**c**) Circuit diagram of the FM measurement setup. (**d**) Current-vs-voltage plot of the device with an external resistor of 58 kΩ. (**e**) Schematic illustration of the frequency modulation detection scheme

processes in the oscillation. The blue process indicates the charging of capacitor C_1 when the circuit is first initiated. During the blue process, the current I_d ignores the VO_2 channel (Rv) due to its high resistance in the insulating phase, and charges the capacitor C_1. Following the charging of C_1 (shaded as the blue process in Fig. 3.3c), the discharging of C_1 (shaded as the red process in Fig. 3.3c) starts as soon as the voltage V_1 across C_1 exceeds a certain threshold to drive sufficient current through Rv to induce an IMT due to the increase in local Joule heating [26–28]. After the IMT, the PCM (Rv) transits to a metallic phase with considerably lower resistance compared to its insulating phase, resulting in the discharging of capacitor C_1. Subsequently, the voltage across C_1 (V_1) decreases, eventually leading to an MIT in the PCM due to the significant decrease in the current/Joule heating.

The continuous repetition of these two processes results in the oscillation in current/voltage across the PCM. One complete electrical oscillation cycle includes one rise and one fall of the voltage V_1 across the PCM (R_V). The rise and fall time can be understood as the time required to travel between two threshold voltages V_{th1} and V_{th2}, where V_{th1} is the threshold voltage at the end of the C_1 charging ($V_1 > V_{th1}$) when the IMT occurs and V_{th2} is the threshold voltage at the end of C_1 discharging ($V_1 < V_{th2}$) when the MIT happens. To acquire the full hysteresis curve of the VO_2 phase transition, we measure the change in the voltage across the VO_2 film by varying the bias current. An external resistor R_{ext} = 58 kΩ is added to the circuit (Fig. 3.3c) to disable electrical oscillation while varying the applied current.

Figure 3.3d plots the measured voltage across the PCM as a function of the bias current. The sudden drop in the voltage in the red curve while increasing the applied current represents IMT, whereas the sudden bump in the voltage while decreasing the applied current represents MIT. The voltages across the PCM (V_1) at the transition edges are recorded as the threshold voltages V_{th1} and V_{th2}, respectively. During the LWIR exposure of the device, the stack absorbs the light and modifies the phase of PCM toward metallic phase. Although this modification is not strong enough to induce the phase transition due to the low intensity of the incident light, it reduces both threshold voltages V_{th1} near IMT and V_{th2} near MIT resulting in the shift in the oscillation frequency. Figure 3.3e shows a schematic illustration of the working principle of this proposed FM-based detection scheme. At first, during dark conditions, the circuit oscillates at a frequency of f_{dark}. When the detector is illuminated with infrared light, it changes the oscillation frequency to $f_{dark} + \Delta f$. Once the incident light is removed, the oscillation frequency returns to its original frequency.

With light illumination, the phase transition property change of VO_2 film is shown in Fig. 3.4. The change in the I–V characteristics leads to the change in the oscillation frequency of the circuit.

Therefore, by recording the frequency shift Δf, the incoming infrared light can be detected. The absorption spectra of the PCM/Si_3N_4 stack in the insulating (blue) and metallic (red) phases are measured and plotted in Fig. 3.5a where a significant change in absorption between the two end states is observed. To corroborate the experimental data with numerical calculations, we perform finite difference time domain (FDTD) simulations in Ansys Lumericals, where there is a good agreement between the simulation (dashed) and experimental results (solid) (see Methods section for details of the FDTD simulation). We consider the absorption of the stack during oscillation to be in between the metallic and insulating phase absorptions. Figure 3.5b shows two measured electrical oscillation waveforms in dark (blue) and light (red) conditions, with the time between the two waveforms adjusted to best present the difference in oscillation frequency. The corresponding oscillation frequencies in the light (red) and dark (blue) conditions are displayed in Fig. 3.5c.

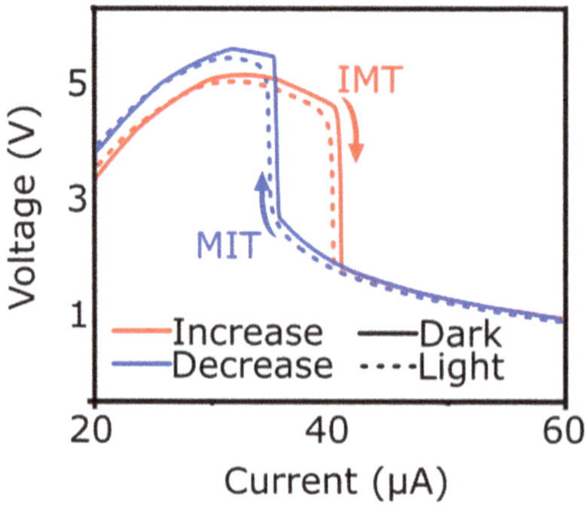

Fig. 3.4 Device I-V curve under dark and light conditions. The hysteresis curves of the VO_2 film are different under light and dark conditions. The transition time from one to another is found different via the rise and fall time measurement of the device

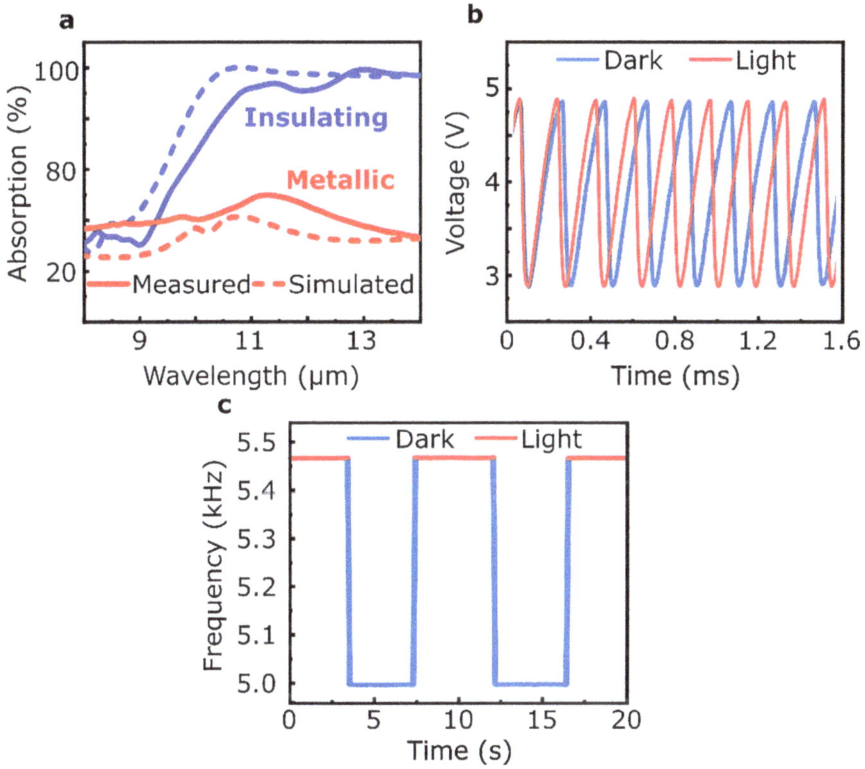

Fig. 3.5 Absorption and measured FM. (**a**) Measured and simulated absorption spectra of the detector in the wavelength range of 8–14 μm in insulating (Blue) and metallic (Red) phase. (**b**) Oscillation waveforms during the dark (Blue) and light (Red) conditions. A higher oscillation frequency is observed when the detector is illuminated with light, and the corresponding shift in the oscillation frequency is shown in (**c**)

3.3.2 Characterization of the Device

Next, the performance of the detector is assessed in terms of pixel dimensions. For this study, we characterized detectors of three different active areas: 8 × 8 μm^2, 12 × 12 μm^2, and 15 × 15 μm^2. Figure 3.6 shows the experimental setup.

Electrical measurements: The fabricated device was driven by a constant current source with a source measurement unit (Keithley 2614B). No capacitor was attached, but the total capacitance (6 nF) of the system was extracted by fitting the experimental data with the simulation. The device was mounted on a probe station (MPI TS150). The oscillation waveform of the voltage across the VO_2 film was recorded by a digital storage oscilloscope (Tektronix Mix Domain Oscilloscope MDO34).

Optical setup: A quantum cascade laser source (OmniLux from Pranalytica, Inc.) at 8.7 μm was focused by a parabolic mirror onto the VO_2 film. Laser power at

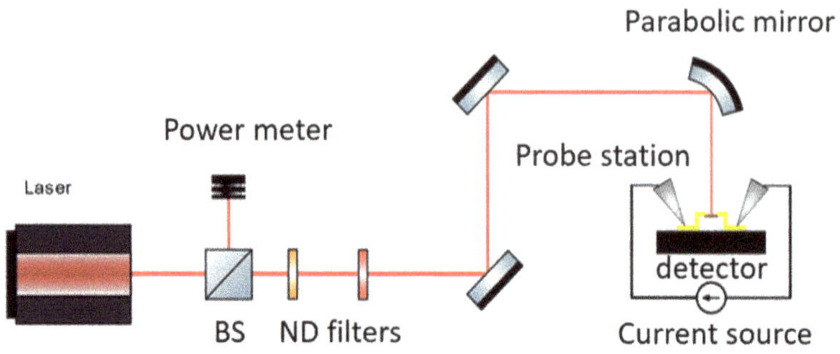

Fig. 3.6 FM-based LWIR detection measurement setup. BS Beam splitter, ND Neutral density

different positions in the beam path was measured by a power meter (Thermal Power Sensor S401C from Thorlabs). Neutral density filters (Thorlabs) were used to scale the incident power. For mapping the beam profile, the parabolic mirror was moved by a motorized XY-stage (M30X, Thorlabs), while the device was kept static. For the single-pixel imaging experiment, a chromium mask on top of ZnSe window (WG70530 from Thorlabs) was placed in between the parabolic mirror and the device. The diameter of the mask plate (ZnSe window) is 2.54 cm, while the size of the dragon pattern on top is 1–2.1 cm. The laser was focused on the mask to achieve the best resolution of the recorded image. Both the parabolic mirror and the sample were kept static, while the mask was moved by the motorized XY-stage (M30X, Thorlabs).

Figure 3.7a shows the evolution of the oscillation frequency as a function of drive current for three different pixel sizes in dark and light conditions. The devices which are smaller in size (8 × 8 μm^2) exhibit the highest oscillation frequency with the same drive current. It is observed that both the limit current I_{min} and I_{max} are larger for the devices with 15 × 15 μm^2 pixel size. Despite the larger operation current, the bigger devices show lower frequency response under the same incident power compared to the smaller devices. The effect of pixel size on the behavior of the devices is attributed to the current density being lower in bigger pixels under the same drive current. The lower current density in the larger pixel leads to less local Joule heating, and hence higher drive currents are required to induce the phase transition in the larger pixels.

Consequently, higher limit currents I_{min} and I_{max} are observed in the case of larger pixels. The decrease in the oscillation frequency with the bigger pixels can be explained by the longer time required to travel between two threshold voltages V_{th1} and V_{th2}, due to the larger difference between them $\Delta V = V_{th1} - V_{th2}$; further details can be found in Fig. 3.8a, b.

Refer to Figs. 3.7a and 3.8b which shows the corresponding frequency shifts of three different pixel-size detectors. While the device with the dimension of 8 × 8 μm^2 exhibits the largest frequency shift, all three sizes show a similar trend as a function of drive current. From the frequency shift, the responsivity R can be determined as follows:

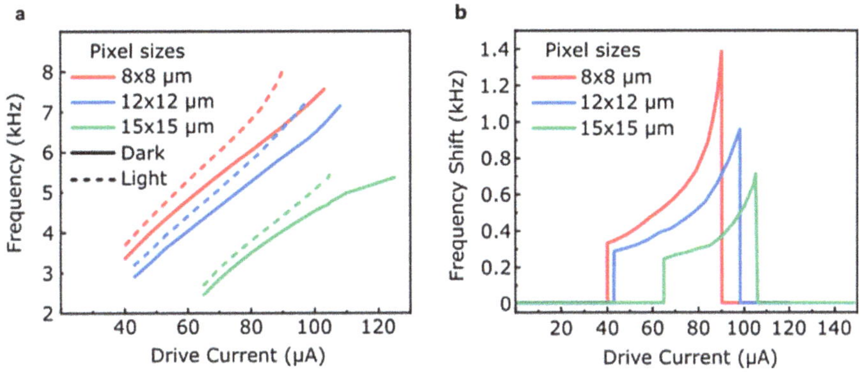

Fig. 3.7 Detector LWIR performance measurements at room temperature. (**a**) Evolution of oscillation frequency of the detector with respect to the driving current for three different sizes of the VO_2 channel; the corresponding frequency shift as a function of drive current is shown in (**b**). The shift is presented as 0 when the circuit is not oscillating

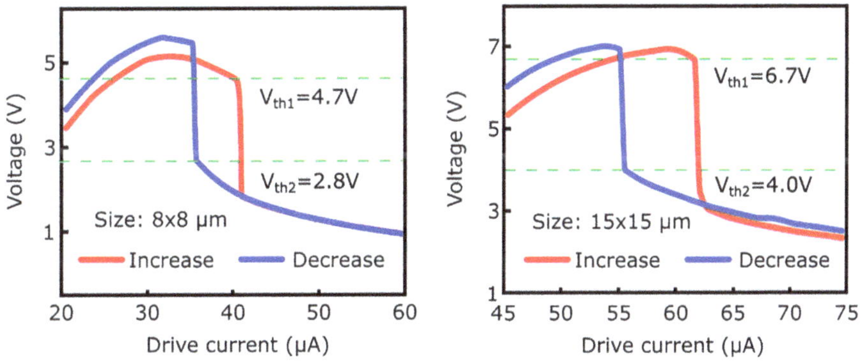

Fig. 3.8 Comparison between different pixel sizes. The larger device exhibits higher $V_{th1}–V_{th2}$ leading to a slower oscillation at the same drive current

$$\mathcal{R} = V_{\mathrm{PTE}} / \mathrm{P}_{\mathrm{inc}} \tag{3.1}$$

where Pinc is the incident power on the active region of the device. It is to be noted that since the spot diameter of the laser is ~300 μm, Pinc is only a fraction of the total incident power. Therefore, from these measurements, it can be inferred that better responsivity is expected for devices with smaller pixel sizes. Further, the sensitivity of the device with 8 × 8 μm^2 active pixel area is assessed by measuring the detectivity (D*) as a function of drive current. The *D** is defined by the expression.

$$D^* = \sqrt{A} / \mathrm{NEP}f \tag{3.2}$$

where A is the active area of the detector and NEP is the noise equivalent power. NEP can be determined by:

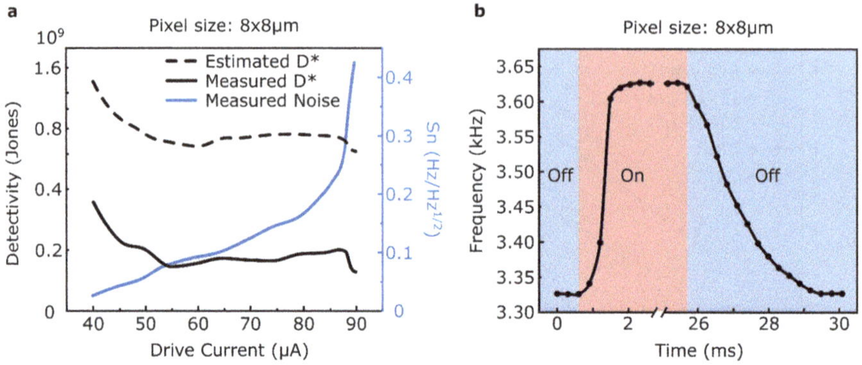

Fig. 3.9 Detectivity and response time. (**a**) Dependence of detectivity and noise of the detector on the drive current for an 8 × 8 μm² pixel. (**b**) Measured time response of the 8 × 8 μm² device with the light switched on (red) and off (blue)

$$\mathrm{NEP} = S_{nf} / \mathcal{R} \tag{3.3}$$

where S_n is the noise spectral density calculated at 1 Hz by a waveform measurement with Shannon-Nyquist sampling theorem. The noise is attributed to circuit FM noises such as phase noise and jitter noise [29, 30].

The evolution of the noise and detectivity as a function of the drive current is plotted in Fig. 3.9a. As the drive current increases, the noise (S_n) also increases because the device approaches the metallic phase. This is attributed to the increased instability of V_{th1} and V_{th1}–V_{th2} with a high drive current. The black graph in Fig. 3.9b displays the measured D^* of the device as a function of the drive current. This measurement demonstrates that when the drive current is set to the smallest value of 40 μA, producing the lowest noise S_n, the highest measured D^* of 3.8×10^8 Jones is obtained for these fabricated detectors. As the drive current increases, the value of Sn also increases, leading to a corresponding decrease in D^*. This decrease in D^* continues until a certain drive current is attained, after which it remains mostly constant. The capacitance of the measurement setup is estimated to be ~6 nF by matching experimental result with the equivalent circuit model prediction. It is worth noting that if the capacitance in parallel with the VO_2 film is reduced or the electrical circuit's instability is decreased, the sensitivity of the proposed FM-based detector for room temperature use will be further enhanced. According to industry standards, constructing a measurement setup with a parallel capacitance of 1 nF is a straightforward task. Therefore, the D^* of the device with 1 nF capacitance is predicted by the equivalent circuit and represented by the red curve in Fig. 3.9a, demonstrating a four times improvement the D^* value is achievable. For a drive current of 40 μA, it is anticipated that the highest expected D^* is around 1.5×10^9 Jones with room temperature detection, which is a level achievable only with a cryogenically cooled LWIR detector among commercially available detectors. The time response is then studied as shown in Fig. 3.9b. The light source is turned on/off with the on/off of the control current of the QCL. The calculated response time τ_{total} is ~2.96 ms at 40 μA by the rise time τ_{rise} and fall time τ_{fall} by $\tau_{total} = \tau_{rise} + \tau_{fall}$. τ_{rise}/τ_{fall} is calculated by the duration between 10% and 90% of changes in steady-state frequencies [31]. The result is five to seven times faster compared to the commercially available uncooled LWIR detectors in the market [32–34]. The short response time of the proposed detector can be attributed to the strong electrothermal feedback in VO_2 thin film [35, 36].

Next, the dependency of the frequency response of the device to the incident IR power is investigated. As discussed previously, the spot size of the excitation laser, with a wavelength of 8.7 μm, is much larger than the active area of the detector, which is 8 × 8 μm^2. Therefore, an accurate estimation of the incident power on the active area of the detector is necessary. For this purpose, we obtain the beam profile of the incident laser in form of a frequency shift map by scanning the detector within the focal plane of the incident laser. The parabolic mirror used to focus the laser is shifted in increments of 10 μm with a motorized XY stage to scan the detector within the laser spot. Figure 3.10a is the measured frequency shift map of the perfectly Gaussian beam profile of the incident laser. The incident power Pinc on the active area of the detector is finally calculated by picking out the center 8x8 μm2 region of the measured frequency shift map as:

$$P_{inc} = \left(\Delta f_d / \Delta f_{map}\right) \times P_{map} \tag{3.4}$$

where Δf_d is the integrated frequency shift over the center 8 × 8 μm^2 region of the map; Δf_{map} is the integrated frequency shift of the whole map; and P_{map} is the total incident power which is measured in front of the parabolic mirror by a power meter. The same measurements are performed to determine the incident laser power for the estimation of R, NEP, and D*. Figure 3.10b shows the shift in oscillation frequency

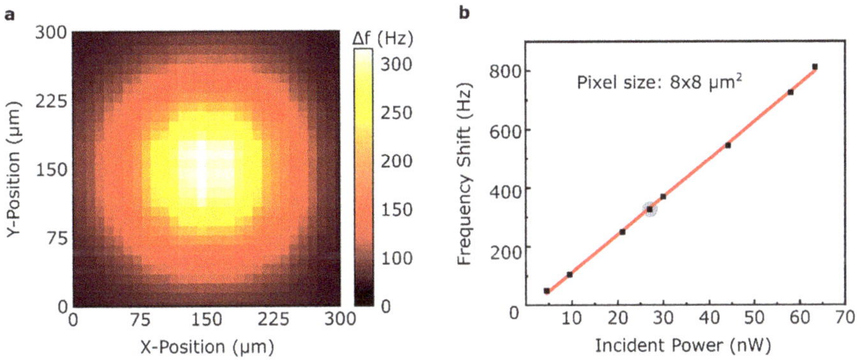

Fig. 3.10 Beam profile imaging and linearity of the photo-response. (**a**) Frequency shift map of the laser beam profile obtained by scanning an 8 × 8 um^2 size pixel. (**b**) Dependence of the frequency shift on the incident power showing a linear detector response

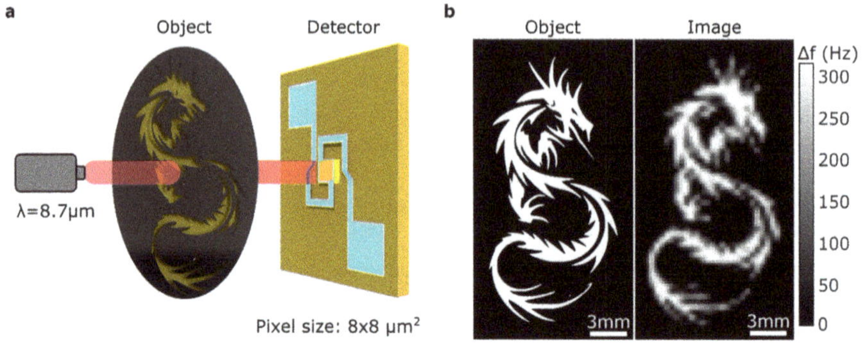

Fig. 3.11 FM-based infrared single pixel imaging. (**a**) Schematic illustration of the experimental setup of the single-pixel imaging system. The light source and detector are fixed while the object is mounted on the motorized XY-stage for scanning. (**b**) Original design of the object (left) and the obtained image (right) through single-pixel imaging in terms of frequency shift

Δf as a function of incident laser power for an $8 \times 8\ \mu m^2$ device. The power of the laser is varied using neutral density filters before the parabolic mirror. A linear response is observed for a large range of incident power on the $8 \times 8\ \mu m^2$ area from 4.54 to 63.4 nW. The device's position relative to the beam is carefully adjusted and placed at the center of the beam by scanning the beam profile for each incident power at the start of the measurement.

To demonstrate the performance of our LWIR detector, a single-pixel imaging experiment was conducted. The setup for the experiment is illustrated in Fig. 3.11a. The QCL laser is sent through a hollow mask of a dragon onto the detector. By changing the position of the mask, we were able to produce the image of the dragon as shown in Fig. 3.11b. It is to be noted that the captured image is a map of frequency shift as opposed to amplitude shift in conventional imaging techniques. Sharp high-resolution features are clearly visible in the captured image. We believe this is the first demonstration of FM-based LWIR imaging.

3.4 Equivalent Circuit Model in LT-Spice

While extensive research has been done on the phase transition of VO_2, the physics behind the phase transition when it is simultaneously driven by current and electric field (in this case incident infrared light) is not well understood. A good model that can describe the full transition is still being explored. For this reason, it is daunting to simulate the oscillation frequency and light-induced modulation of the frequency of our device at the same time for a wide range of currents, because this would require describing the phase transition both under DC (bias current) and high-frequency (incident IR radiation) electric fields. We instead followed an approximate approach by creating an equivalent circuit of the device in a circuit simulation

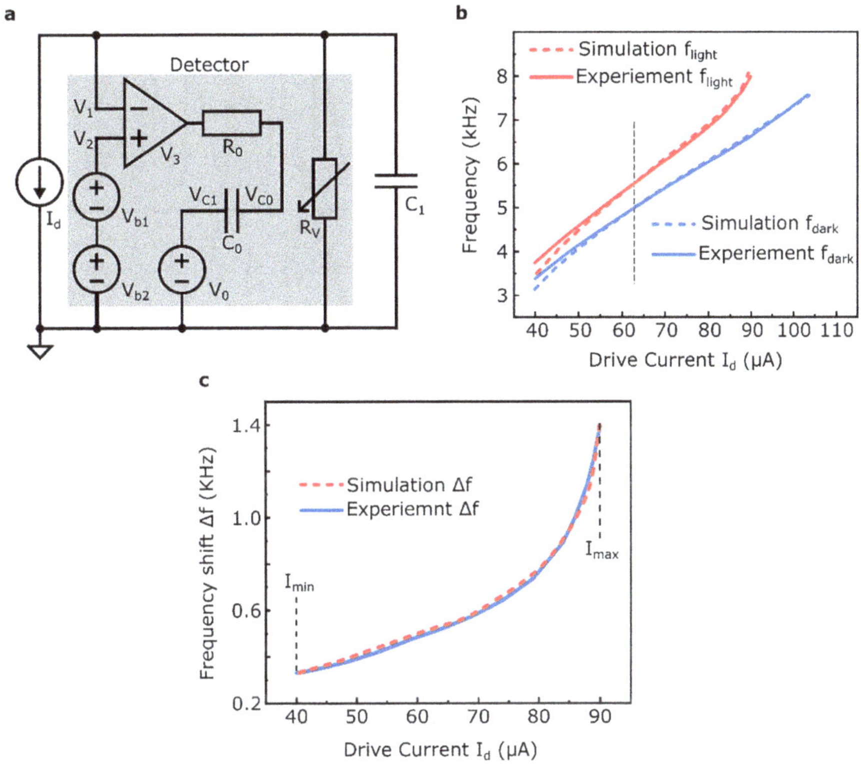

Fig. 3.12 FM-based detector equivalent circuit model. (**a**) Equivalent circuit diagram of the FM-based detector in LTspice. (**b**) Evolution of the frequency of oscillation as a function of drive current Id for an 8 × 8 μm² active detector area under dark (blue) and light (red) conditions. (**c**) Simulated and experimental frequency shift Δf between light and dark conditions as a function of the drive current of the oscillator

environment to match and predict the electrical performance of the device. To simulate the oscillation frequency and its light-induced modulation, we have created an equivalent circuit model of the device in LTspice [37, 38], which is illustrated in Fig. 3.12a. In this model, constant current source Id is the drive current used in the experiment. R_V is a voltage-controlled resistor representing the resistance of the VO_2 film and C_1 refers to the capacitance in the system that is connected in parallel to the VO_2 film. It should be noted that Id and C_1 are among the external components utilized in the experiment. Additional circuit elements including a voltage comparator (V_3), constant voltage sources (V_{b1}, V_{b2}, V_0), a constant resistor (R_0), and a capacitor (C_0) shown with a gray background, along with R_V, are incorporated to imitate the variation of the PCM's voltage-controlled resistance (R_V) during phase transition. Details of the simulation are included in the Methods section. Figure 3.12b plots the oscillation frequency of the circuit as a function of drive current during the dark conditions (blue) and when illuminated with an 8.7 μm wavelength IR laser (red).

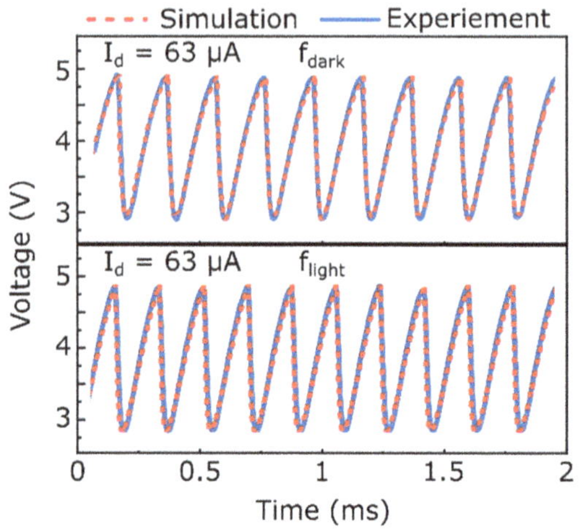

Fig. 3.13 Waveform simulation with the equivalent circuit model. Simulated and experimental frequency shift Δf between light and dark conditions as a function of the drive current of the oscillator

An increase in the oscillation frequency is observed during light illumination compared to the dark condition. The simulation agrees well with the experimental results. It should be noted that the circuit can support oscillations in a certain range of the drive current. For the devices with 8 × 8 μm^2 active area, oscillation onsets from the drive current at 40 μA and ceases at 108 μA in the dark condition. When the drive current is lower than 40 μA, the voltage $V_1 < V_{th1}$ on the VO_2 film stays below the IMT threshold and consequently no current oscillation exists.

On the other hand, with a high drive current above 108 μA, the device makes an IMT but fails to make an MIT afterward due to $V_1 > V_{th2}$ at all the time, and hence the oscillation cannot exist as well. Here, I_{min} and I_{max} are referred to as the minimum and maximum limit currents that support oscillation in the circuit, respectively. The shift in the frequency in presence of light ($\Delta f = f_{light} - f_{dark}$) as a function of drive current for a constant illumination power is shown in Fig. 3.12c. It is observed that Δf increases with the drive current I_d. As the drive current Id approaches the threshold current I_{th2}, the increase in Δf is particularly rapid.

Figure 3.13 displays the corresponding simulated (red dashed line) and experimental (blue solid line) oscillation waveforms at 63 μA in the dark (top) and the light (bottom) conditions. The experiments demonstrate that the frequency of the circuit increases from 5.007 to 5.506 kHz when exposed to light, manifesting a frequency shift of $\Delta f \sim 500$ Hz which shows an excellent match with the simulation results.

3.5 Discussion

In summary, we have demonstrated an FM-based uncooled LWIR detector using a cavity-coupled phase-change material. The FM-based detection scheme possesses inherent robustness against noise compared to conventional AM-based detection schemes, providing significant improvement in sensitivity and signal-to-noise ratio while minimizing the influence of external interference. The measured detectivity of the proposed detector is higher than commercially available uncooled LWIR detectors with a faster response time. We expect that a proper industrial packaging with low capacitance circuit components can further increase the detectivity comparable to cryogenically cooled detectors (~10^9). Notably, the detector exhibits a highly desirable linear response and high detectivity at room temperature, rendering it a promising solution for diverse applications. Furthermore, we have demonstrated the feasibility of developing a multipixel IR camera with these detectors through single-pixel imaging. Although we employed VO_2 as the phase-change material in our work, this detection scheme can be extended to other phase-change materials exhibiting NDR, and different spectral regions of light can be covered depending on the absorption properties of the material. Furthermore, PCMs such as W-doped VO_2 with lower phase transition temperature [39] can be operated with smaller drive current, which is beneficial for the focal plane array integration. The operational wavelength may also be tuned by changing the cavity thickness. Our results introduce this novel FM-based detector as a unique platform for creating low-cost, high--efficiency uncooled infrared detectors and imaging systems for various applications such as remote sensing, thermal imaging, and medical diagnostics.

References

1. Vincent, J. D., Hodges, S., Vampola, J., Stegall, M., & Pierce, G. (2015). *Fundamentals of infrared and visible detector operation and testing*. Wiley.
2. Yadav, P. V. K., et al. (2022). Advancements of uncooled infrared microbolometer materials: A review. *Sensors and Actuators A: Physical, 342*, 113611. https://doi.org/10.1016/j.sna.2022.113611
3. Donati, S. (2021). *Photodetectors: Devices, circuits and applications*. Wiley.
4. Li, Z., et al. (2017). Imaging metal-like monoclinic phase stabilized by surface coordination effect in vanadium dioxide nanobeam. *Nature Communications, 8*, 15561.
5. Wentzcovitch, R. M., Schulz, W. W., & Allen, P. B. (1994). VO_2: Peierls or Mott-Hubbard? A view from band theory. *Physical Review Letters, 72*, 3389–3392. https://doi.org/10.1103/PhysRevLett.72.3389
6. Guo, T., et al. (2023). Frequency modulation based long-wave infrared detection and imaging at room temperature. *Advanced Functional Materials*, 2309298.
7. Ke, Y., et al. (2018). Vanadium dioxide: The multistimuli responsive material and its applications. *Small, 14*, 1802025.
8. Goodenough, J. B. (1971). Metallic oxides. *Progress in Solid State Chemistry, 5*, 145–399.
9. Schwingenschlögl, U., & Eyert, V. (2004). The vanadium Magnéli phases VnO2n-1. *Annalen der Physik, 13*, 475–510.
10. Sasaki, T., Ueda, H., Kanki, T., & Tanaka, H. (2015). Electrochemical gating-induced reversible and drastic resistance switching in VO2 nanowires. *Scientific Reports, 5*, 1–7.
11. Liu, K., Lee, S., Yang, S., Delaire, O., & Wu, J. (2018). Recent progresses on physics and applications of vanadium dioxide. *Materials Today, 21*, 875–896. https://doi.org/10.1016/j.mattod.2018.03.029

12. Shao, Z., Cao, X., Luo, H., & Jin, P. (2018). Recent progress in the phase-transition mechanism and modulation of vanadium dioxide materials. *NPG Asia Materials, 10*, 581–605. https://doi.org/10.1038/s41427-018-0061-2
13. Lee, M. J., et al. (2007). Two series oxide resistors applicable to high speed and high density nonvolatile memory. *Advanced Materials, 19*, 3919–3923.
14. Crunteanu, A., et al. (2010). Voltage- and current-activated metal–insulator transition in VO2--based electrical switches: A lifetime operation analysis. *Science and Technology of Advanced Materials, 11*, 065002. https://doi.org/10.1088/1468-6996/11/6/065002
15. Xu, X., et al. (2012). The extremely narrow hysteresis width of phase transition in nanocrystalline VO2 thin films with the flake grain structures. *Applied Surface Science, 261*, 83–87. https://doi.org/10.1016/j.apsusc.2012.07.098
16. Nandi, S. K., Das, S. K., Estherby, C., Gentle, A., & Elliman, R. G. (2020). Understanding modes of negative differential resistance in amorphous and polycrystalline vanadium oxides. *Journal of Applied Physics, 128*, 244103. https://doi.org/10.1063/5.0027875
17. Corti, E., et al. (2020). Scaled resistively-coupled VO2 oscillators for neuromorphic computing. *Solid-State Electronics, 168*, 107729. https://doi.org/10.1016/j.sse.2019.107729
18. Shukla, N., et al. (2014). Synchronized charge oscillations in correlated electron systems. *Scientific Reports, 4*, 4964. https://doi.org/10.1038/srep04964
19. Karaoglan-Bebek, G., Hoque, M. N. F., Holtz, M., Fan, Z., & Bernussi, A. A. (2014). Continuous tuning of W-doped VO2 optical properties for terahertz analog applications. *Applied Physics Letters, 105*. https://doi.org/10.1063/1.4902056
20. Parihar, A., Shukla, N., Jerry, M., Datta, S., & Raychowdhury, A. (2017). Computing with dynamical systems based on insulator-metal-transition oscillators. *Nano, 6*, 601–611. https://doi.org/10.1515/nanoph-2016-0144
21. Salverda, M., Hamming-Green, R. P., & Noheda, B. (2022). An epitaxial perovskite as a compact neuristor: Electrical self-oscillations in TbMnO3 thin films. *Journal of Physics D: Applied Physics, 55*, 335305. https://doi.org/10.1088/1361-6463/ac71e2
22. Liu, X., Li, S., Nandi, S. K., Venkatachalam, D. K., & Elliman, R. G. (2016). Threshold switching and electrical self-oscillation in niobium oxide films. *Journal of Applied Physics, 120*. https://doi.org/10.1063/1.4963288
23. Das, S. K., et al. (2023). Physical origin of negative differential resistance in V3O5 and its application as a solid-state oscillator. *Advanced Materials, 35*, 2208477. https://doi.org/10.1002/adma.202208477
24. Pattanayak, M., Hoque, M. N. F., Fan, Z., & Bernussi, A. A. (2018). Electrical oscillation generation with current-induced resistivity switching in VO2 micro-channel devices. *Science and Technology of Advanced Materials, 19*, 693–701. https://doi.org/10.1080/14686996.2018.1521249
25. Safaei, A., et al. (2017). Dynamically tunable extraordinary light absorption in monolayer graphene. *Physical Review B, 96*, 165431. https://doi.org/10.1103/PhysRevB.96.165431
26. Zimmers, A., et al. (2013). Role of thermal heating on the voltage induced insulator--metal transition in VO2. *Physical Review Letters, 110*, 056601. https://doi.org/10.1103/PhysRevLett.110.056601
27. Kim, H.-T., et al. (2005). Raman study of electric-field-induced first-order metal-insulator transition in VO2-based devices. *Applied Physics Letters, 86*. https://doi.org/10.1063/1.1941478
28. Oka, T., Arita, R., & Aoki, H. (2003). Breakdown of a Mott insulator: A nonadiabatic tunneling mechanism. *Physical Review Letters, 91*, 066406. https://doi.org/10.1103/PhysRevLett.91.066406
29. Hajimiri, A., Limotyrakis, S., & Lee, T. H. (1999). Jitter and phase noise in ring oscillators. *IEEE Journal of Solid-State Circuits, 34*, 790–804.
30. Lax, M. C., & noise. V. (1967). Noise in self-sustained oscillators. *Physical Review, 160*, 290.
31. Zhang, M., et al. (2018). High-performance photodiode-type photodetectors based on polycrystalline formamidinium Lead iodide perovskite thin films. *Scientific Reports, 8*, 11157. https://doi.org/10.1038/s41598-018-29147-6

32. Richards, P. L. (1994). Bolometers for infrared and millimeter waves. *Journal of Applied Physics, 76*, 1–24. https://doi.org/10.1063/1.357128
33. Li, C., Han, C.-J., & Skidmore, G. (2011). Overview of DRS uncooled VOx infrared detector development. *Optical Engineering, 50*, 061017-061017-061017.
34. Schimert, T. et al. (2009). *Infrared technology and applications XXXV.* 289–293 (SPIE).
35. Neto, G., et al. (2008). Figures of merit and optimization of a VO2 microbolometer with strong electrothermal feedback. *Optical Engineering, 47*, 073603-073603-073615.
36. Irwin, K. (1995). An application of electrothermal feedback for high resolution cryogenic particle detection. *Applied Physics Letters, 66*, 1998–2000.
37. Maffezzoni, P., Daniel, L., Shukla, N., Datta, S., & Raychowdhury, A. (2015). Modeling and simulation of vanadium dioxide relaxation oscillators. *IEEE Transactions on Circuits and Systems I: Regular Papers, 62*, 2207–2215. https://doi.org/10.1109/TCSI.2015.2452332
38. Amer, S., Hasan, M. S., Adnan, M. M., & Rose, G. S. (2019). SPICE modeling of insulator metal transition: Model of the critical temperature. *IEEE Journal of the Electron Devices Society, 7*, 18–25. https://doi.org/10.1109/JEDS.2018.2875627
39. Sun, K., et al. (2022). Room temperature phase transition of W-doped VO2 by atomic layer deposition on 200 mm Si wafers and flexible substrates. *Advanced Optical Materials, 10*, 2201326.

Chapter 4
Dense Pixel Array Integration

Abstract Although researchers have developed numerous impressive tools within the confines of research laboratories, scaling them up for industrial applications poses a significant challenge. When it comes to LWIR detectors, most of the reported work focuses on individual pixels, with limited representation of innovative detectors featuring dense pixel arrays and integrated readout circuits (ROIC) (Schultz et al. Lincoln Lab J 20:36–51, 2014; Zverev et al. J Phys Conf Ser. IOP Publishing). In this presentation, I will showcase our endeavors to integrate the graphene LWIR detector into a commercially available ROIC with dense pixel focal plane array, paving way to prototype camera production.

Please note that, despite the tremendous time and effort invested, due to intellectual property protection and restrictions, this aspect of my work will be presented in a concise and descriptive manner. Figure below illustrates an example of a readout circuit from an open access source (Zverev et al. J Phys Conf Ser. IOP Publishing) (not the ROIC used in this work).

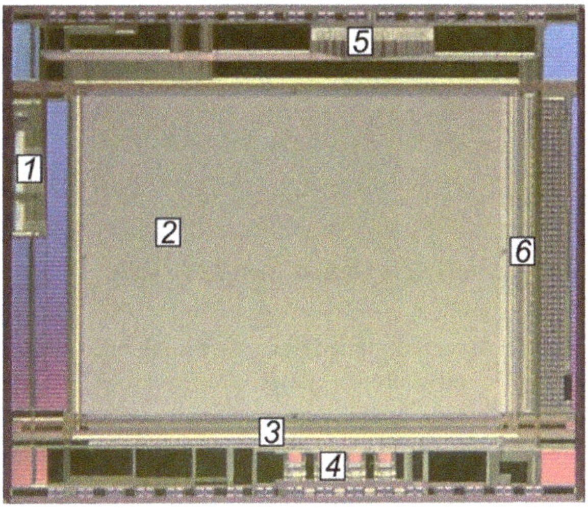

T. Guo, *Low Energy Photon Detection*, Springer Theses,
https://doi.org/10.1007/978-3-031-71544-0_4

Example of an electronic readout circuit die photo (Zverev et al. J Phys Conf Ser. IOP Publishing)
1 – DACs block, 2 – array of pixel cells, 3 – column followers and decoders, 4 – preamplifiers and output buffers, 5 – digital block, 6 – row decoder

Focal plane array (FPA) is a collection of light-sensitive pixels arranged in a typically rectangular grid at the focal plane of a lens. FPAs are primarily employed for imaging tasks, such as capturing photographs or video footage. The FPA lies in the ROIC which is labeled as 2 in figure above. With FPA and the imaging lens array, LWIR images can be generated at the output of ROIC as shown in figure below (Zverev et al. J Phys Conf Ser. IOP Publishing).

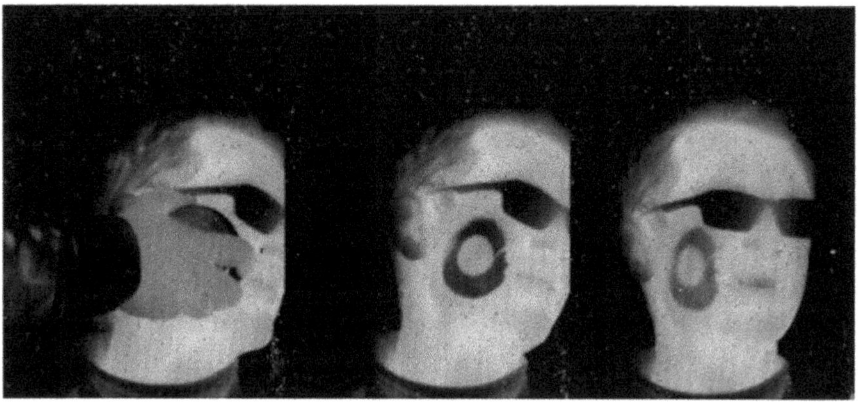

Heat residue on a person's face imaged with FPA (Zverev et al. J Phys Conf Ser. IOP Publishing)

4.1 Experimental Method

4.1.1 Fabrication of the Specially Designed Single Pixel

The graphene-based single pixel device is redesigned based on the configuration of the pixels on the ROIC. Instead of a three-terminal device, we have changed our design to a two-terminal device with source and drain electrodes connecting the graphene. Instead of electrostatic doping, we employed chemical doping to increase the carrier concentration in graphene.

4.1.2 Test of Performance of the Single Pixels

The performance of the single pixel device is tested by a probe station after all fabrication devices. And we could get decent detectivity with the alternated design.

4.1.3 Fabrication of Dense Pixel Array on ROIC

Fabricating the dense pixel array on the ROIC presented a formidable challenge for several reasons. First, a different material for the optical cavity is required. Fortunately, a near-transparent cavity was successfully optimized and fabricated. Secondly, working on a small die as opposed to an entire wafer posed a different set of challenges from multiple perspectives. Finally, the electronics on the ROIC and creating additional pixels with various functionalities also proved to be highly demanding tasks.

Despite the numerous challenges encountered, the integration ultimately proved to be a success. Figure 4.1a, b demonstrates the successful integration of the LWIR graphene detector onto the ROIC substrate.

4.1.4 Characterization of the Dense Pixel Array

The fabricated detector is first examined with an optical microscope and SEM to ensure the critical element is present. Then the variation between different graphene pixels is carefully examined as excess difference on the resistance with active pixels will cause issue in calibration. A series of tests on other elements on ROIC is

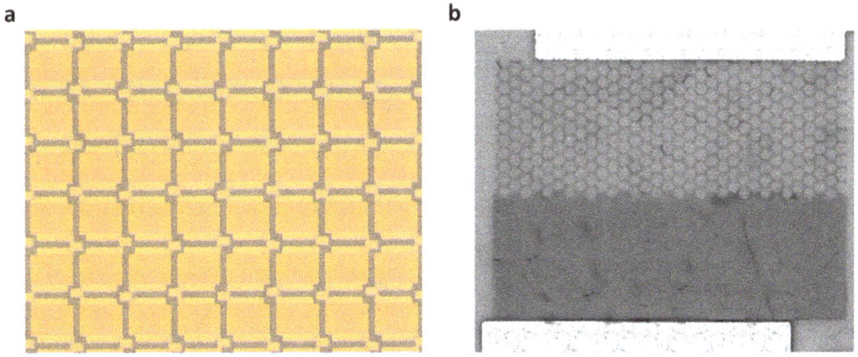

Fig. 4.1 Device size vs detectivity vs estimated response time. (**a**) Microscope image of fabricated LWIR detector on ROIC. (**b**) SEM image of a single pixel of the fabricated detector on ROIC

performed as well. These tests validate the ability of the fabricated focal plane array's functionality on the ROIC electronically. Finally, the optical and advanced electronical test is ongoing with an industry collaborator.

4.2 Discussion

I've showcased the design and performance verification of single-pixel LWIR detectors constructed from graphene, employing a unique design that ensures compatibility with the ROIC. Following this achievement, the optimization of fabrication steps for integrating a dense pixel array was successfully completed. It's essential to acknowledge that fabricating a dense pixel array for graphene devices presents a significantly higher level of complexity compared to crafting a single pixel, and there remains room for improvement. Nevertheless, through thorough scrutiny using various tools, we have confirmed the successful formation of the dense pixel array. The fabricated dense pixel array detector is undergoing performance evaluation in collaboration with a partner.

References

1. Schultz, K. I., et al. (2014). Digital-pixel focal plane array technology. *Lincoln Laboratory Journal, 20*, 36–51.
2. Zverev, A. V. et al. (2015). 384x288 readout integrated circuit for MWIR and LWIR HgCdTe based FPA. In *Journal of Physics: Conference Series* (Vol. 643, No. 1). IOP Publishing.

Chapter 5
Conclusion and Future

Abstract In this chapter, I will conclude the thesis by comparing the graphene-based and VO_2-based detector with traditional infrared detectors to demonstrate their advantages. Furthermore, I will present a few future directions for both work and the possible outcomes. At last, the integration of the graphene detector to the ROIC could pave way to potential successful commercialization.

5.1 Conclusion

In conclusion, I have worked on two approaches of LWIR detectors with room-temperature operation. Both detectors turned out yielding better detectivity compared to traditional microbolometers with a faster response time. The graphene-based LWIR detector featured in this work relies on Dirac plasmon and the Seebeck effect to generate a photo-thermoelectric voltage, while the VO_2-based detector operates as a thermal sensor, detecting local joule heat changes induced by IR illumination.

Both varieties of LWIR detectors hold immense promise for the future. In the case of the graphene LWIR detector described in this study, it's worth noting that only half of the graphene film is photoactive. In the future, a fully decorated graphene film could potentially operate in a similar manner with asymmetric hot carrier generation, while boosting the sensitivity of the detector. Furthermore, the integration of the graphene LWIR detector onto a dense pixel array is an exciting project. The process of integrating it with the ROIC could yield game-changing results and pique commercial interest.

The FM-based detection is the first demonstration of its kind in the LWIR region. The VO_2 used in the work can be changed to other PCMs exhibiting NDR property. As different PMCs have different properties such as absorptions, thermal coefficients, and current/voltage response, integrating them into the FM-based detection scheme will inherently modify the detector, including different absorption band, faster response time, polarization dependence, and opto-mechanical effect.

T. Guo, *Low Energy Photon Detection*, Springer Theses,
https://doi.org/10.1007/978-3-031-71544-0_5

Curriculum Vitae

Tianyi Guo

Dr. Tianyi Guo received his PhD degree in December 2023 from the University of Central Florida. He is currently working as a post-doctoral researcher at University of Central Florida. He was awarded with the Springer These Award in 2024. His research mainly focuses on infrared detection and related nanotechnologies with 2D materials, phase change materials, and semiconductors.

(+1) 407-701-7345 • tianyiguo63@gmail.com

Summary

Optical Physicist in optics, applied physical sciences and engineering. Project experience includes (1) high power femtosecond laser and attosecond pulses generation; (2) the simulation, design, fabrication, experiment and characterization of opto-electronic devices based on phase change materials, 2D materials and semiconductors.

Project Experience

University of Central Florida Post-doctoral Fellow

- Worked on long wave infrared detection based on different materials
- Worked on THz spectroscopy for gas detection
- Worked on near field hyperspectral imaging and spectroscopy for nanoplastic locating and identification

T. Guo, *Low Energy Photon Detection*, Springer Theses,
https://doi.org/10.1007/978-3-031-71544-0

University of Central Florida Graduate Research Assistance

- Studied optoelectronic devices and their applications in infrared detection based on plasmon of 2D materials, phase change materials, semiconductors and nanoparticles.
- Experience with integration/fabrication of micro-devices on electronic read-out circuit (ROIC) and dense-pixel focal plane array (FPA).
- Superuser of multiple PVD tools for equipment maintenance, trouble-shooting and user training. Heavy user of cleanroom.
- Studied high-power fs lasers and as pulses generation with a home-built setup.
- Mentoring undergraduate students.
- Working in cross-functional collaborations with multiple groups.

University of Science and Technology of China Undergraduate Student

- Solar cell with none-toxic quantum dots.

Skills Summary

- Skilled in nano-fabrication, experiment, characterization, and numerical optimization of the opto-electronic devices for infrared detection.
- Skilled in customization of optical and electronic experimental setups, modification, maintenance and troubleshooting of instruments.
- Skilled in advanced data analysis and design of experiments for improved efficiency and throughput in fabrication.
- Skilled in failure analysis, recipe optimizations and conduct of related experiments in fabrication.
- Skilled in PVD (E-beam deposition, thermal deposition, sputtering), lithography (E-beam and UV lithography) and etch (RIE/ICP, plasma etching, wet chemical etching) processes.
- Skilled in design, building, alignment, experiment and maintenance of laser systems and other complex optical setups.
- Writing and optimizing standard operating procedures (SOP) and measurement reports.
- Effective scientific communication with good presentation skills.

Techniques and Instrumentation

Fabrication: DC and RF Sputtering, ALD, photo-lithography, laser writer, E-beam lithography, CVD/PECVD, E-beam and thermal deposition, RIE/ICP, wet etching techniques, 2D-material transfer

Experiment: CPA, OPA, AOM, deformable mirror/wavefront sensor for beam shaping, pump-probe spectroscopy, nonlinear harmonic generations, electro-optical measurements, lock-in measurements, optical testing, optical design

Characterization: Raman Spectroscopy and TERS, FTIR, SEM, AFM, s-SNOM, Ellipsometry
Data Analysis and Statistics: Matlab, Origin, Image J, Gwyddion
Design and Modeling Tools: Zemax, LabVIEW, Autocad, LT-Spice, K-layout, L-edit, COMSOL Multiphysics, Lumerical FDTD solutions
Programming: Python, C++

Education

Ph.D., Physics
University of Central Florida, Orlando, FL, USA
College of Science
College of Optics and Photonics (CREOL)
NanoScience and Technology Center

B.S., Physics
University of Science and Technology of China, Hefei, China
Department of Physics (Major: Condensed Matter Physics)

Publications

1. Frequency Modulation based Long-Wave Infrared Detection and Imaging at Room Temperature, Advanced Functional Materials.
2. Dynamic Photoresponse Tuning in Nano-patterned Monolayer Graphene for Fast and Uncooled Infrared Detection, Nano Letters (Under review)
3. Large-scale coloring of colorless mirrors with ultra-thin transparent dielectrics, ACS Photonics
4. Reusable Structural Colored Nanostructure for Powerless Temperature and Humidity Sensing, Advanced Optical Materials (2023): 2300300
5. Beam optimization in a 25 TW femtosecond laser system for high harmonic generation, Molecular and Optical Physics 53.14 (2020): 145602.
6. Enhanced high-order harmonic generation driven by a wavefront corrected high-energy laser, Journal of Physics B: Atomic, Molecular and Optical Physics 51.13 (2018): 134005.

Milton Keynes UK
Ingram Content Group UK Ltd.
UKHW022024101024
449422UK00001B/17